PERFORMANCE FUEL INJECTION SYSTEMS

Overview of OEM Systems • Tuning Stock ECUs, Piggback and Standalone Units • Drag Strip and Dyno Tuning Tips • EFI Conversions

Matt Cramer & Jerry Hoffmann

HPBooks

HPBooks

Published by the Penguin Group
Penguin Group (USA) Inc.
375 Hudson Street, New York, New York 10014, USA

Penguin Group (Canada), 90 Eglinton Avenue East, Suite 700, Toronto, Ontario M4P 2Y3, Canada
(a division of Pearson Penguin Canada Inc.)
Penguin Books Ltd., 80 Strand, London WC2R 0RL, England
Penguin Group Ireland, 25 St. Stephen's Green, Dublin 2, Ireland (a division of Penguin Books Ltd.)
Penguin Group (Australia), 250 Camberwell Road, Camberwell, Victoria 3124, Australia
(a division of Pearson Australia Group Pty. Ltd.)
Penguin Books India Pvt. Ltd., 11 Community Centre, Panchsheel Park, New Delhi—110 017, India
Penguin Group (NZ), 67 Apollo Drive, Rosedale, North Shore 0632, New Zealand
(a division of Pearson New Zealand Ltd.)
Penguin Books (South Africa) (Pty.) Ltd., 24 Sturdee Avenue, Rosebank, Johannesburg 2196, South Africa

Penguin Books Ltd., Registered Offices: 80 Strand, London WC2R 0RL, England

While the author has made every effort to provide accurate telephone numbers and Internet addresses at the time of publication, neither the publisher nor the author assumes any responsibility for errors, or for changes that occur after publication. Further, the publisher does not have any control over and does not assume any responsibility for author or third-party websites or their content.

PERFORMANCE FUEL INJECTION SYSTEMS

Copyright © 2010 by Matt Cramer & Jerry Hoffmann
Cover design by Bird Studios
Front photo by Mike Mavrigian
Back cover photos by Matt Cramer & Jerry Hoffmann
Interior photos by authors unless otherwise noted

First edition: August 2010

ISBN: 978-1-55788-557-9

PRINTED IN THE UNITED STATES OF AMERICA

10   9   8   7   6   5   4   3   2

NOTICE: The information in this book is true and complete to the best of our knowledge. All recommendations on parts and procedures are made without any guarantees on the part of the author or the publisher. Tampering with, altering, modifying, or removing any emissions-control device is a violation of federal law. Author and publisher disclaim all liability incurred in connection with the use of this information. We recognize that some words, engine names, model names, and designations mentioned in this book are the property of the trademark holder and are used for identification purposes only. This is not an official publication.

CONTENTS

ACKNOWLEDGMENTS

We all start off as beginners. While there are some lessons we've learned through practice and experience, there are also many lessons we've learned from others, either by working alongside them on a problem, or by picking their brains.

First, we'd like to thank Bruce Bowling and Al Grippo, and many of the members of the MegaSquirt community, who have been so generous in sharing what they know, particularly (in alphabetical order): Jean Belanger, Ken Culver, Eric Fahlgren, Lance Gardiner, Scott Hall, James Laughlin, James Murray, Phil Ringwood, Phil Tobin, and Andy Whittle.

We'd also like to thank EFI tuning wizards Scott "Dieselgeek" Clark, Chris Macellaro, Ed Senf, and Scott Siegel. A big thanks goes to Michael Lutfy, our editor, without whom this book would never have been possible. And of course to our wives Joy Hoffmann and Kelly Cramer, who put up with us and our automotive obsessions day in and day out.

Finally, we'd like to thank all of our customers at DIYAutoTune.com who've supported us over the years. Many times we had to do some careful research and testing to get them the right answers, and over time we realized that somebody needed to write all those answers down in a manner anyone could understand. We hope we have accomplished that and more here, giving our readers the practical know-how and confidence to dive in and make their projects happen.

INTRODUCTION

If you've picked up this book, chances are you're a car nut and a serious DIY (do-it-yourself) enthusiast. The whole automotive DIY movement has returned to fashion in recent years, with a multitude of TV shows, books, magazines, online blogs, and clubs dedicated to the hobby. I have always liked to play with cars. I drive around and see different types of cars—some pretty universally loved, and others complete oddballs most would never consider building something cool out of—and I immediately have a vision for them. Whether it's an old Ford/Chevy/Mopar just begging for a new fire-breathing small block, or a little lightweight four-cylinder car I can lighten further, add a little boost, and turn into a great corner-carving rocket, I'm always looking for new project cars. I've even been taking a hard look at my wife's minivan. I'm pretty sure I can fit a second motor in the rear, turbocharge both motors, and still maintain at least one of the bench seats (after getting her a new one first, of course!). There is no shortage of ideas for what I might do next. I bet many of you have similar visions.

One thing that comes into play with every new motorized toy though, is the need to build an air/fuel induction system that produces maximum performance while maintaining excellent drivability. And maybe, depending on the application's needs, it needs a few more features to help get that perfect launch every time. Or maybe you need to cool the intake charge with a blast of water/meth injection, or inject some nitrous. There are ways to accomplish much of this with a carburetor and a few black boxes, but the peformance automotive aftermarket has come a long way since the days where that was the only option. Computer-controlled fuel injection systems are the norm on anything made from the mid-1980s, and retrofitting EFI to a classic car or truck can enable it to perform better, more economically, and with the best drivability you've ever seen in a classic hot rod. You can ditch the black boxes and have a central point of control for every subsystem on your ride: from basic fuel

and ignition requirements to control of a turbo's boost pressure, a multi-stage nitrous injection (and the fueling/ignition changes to go along with it), a two- or three-step rev limiter, and many other features. All of these can be controlled from just a single interface, minimizing the pieces of equipment—and therefore the points of failure, in a system packed full of black boxes.

In short, this book is actually a new tool, of sorts, and an extremely powerful one at that. In many cases (depending on your choice of EFI system), it will help you assume full control of every aspect of your engine's performance. It will give you the ability to perfectly map fuel and ignition at all load ranges and rpm to extract every bit of horsepower from your motor in all scenarios. This tool is a little more complicated than the screwdriver you may have used to tune carburetors, but it's intensely more powerful, giving you a level of control that screwdriver could never accomplish.

Our purpose for writing this book is to remove some of the anxiety many DIYer's have when entering the world of electronic fuel injection. Our goal is to present the semi-complex topic of EFI in a practical manner that will allow you to install, modify, and tune an EFI system on just about any motor. In our daily business at DIYAutoTune.com, we've helped thousands dive into the world of EFI, most without any prior experience, and emerge with a crazy, fun ride to enjoy. So whether you're an old-school carb-guy looking to learn a little bit about this newfangled computer-controlled stuff, an electronics/computer geek looking for something else to tinker with, or anywhere in between, you'll find this book to be a great EFI tool. From the basics to dialing your motor in with the perfect tune, we cover it from a practical angle that will give you the knowledge and the confidence to dive into your next project, which we hope will be the most rewarding one you'll ever undertake.

—*Jerry Hoffmann*

Chapter 1

An Introduction to EFI Tuning

Tuning theories are useful, but they're a means to an end. That end is maximum horsepower and good drivability.

If you're looking for an engineering textbook full of detailed theories and complicated mathematical equations on electronic fuel injection (EFI), put this one down... you've got us all wrong. But if you're the aspiring tuner looking for someone to help you skip through all the bull, get the skinny on what you need to know without getting bogged in the muck, and get a handle on this EFI thing without making you feel like you need to go back to school for a degree in electrical engineering first, keep reading.

The fact is the math and the in-depth theory is good information, but for most of us, it's not *necessary* information. Getting a fuel-injected engine to run well doesn't require you to know how to write control algorithms any more than changing a head gasket requires knowing how to calculate the maximum strain on the head bolts. As long as you're using a reputable product, you can be fairly certain the designers have done their homework and you will just need to get it installed and adjusted right. You can always go back and read up on serious theory later. Our goal is to give you what you need to make it work, and make it work well, without giving you a headache or requiring you to own a calculator with six bazillion buttons. And while a bit of theory is good, it can be taken too far too quickly—you just want to drive your car!

There will always be some basic math you can't escape from; we can't help you completely escape from all thought and we won't pretend to. Sizing injectors, fuel pumps, etc., are fairly basic principles that we'll cover to make this sort of

thing a snap. But at the same time you really don't need to understand the ideal gas law and associated formulae to install and properly tune your engine management system. So we'll try not to bog you down in the details. We've put most of the heavy math off safely in sidebars that you can skip while you read the main text. For those who like having the math where you can find it in a hurry, we've also included all of the equations in an appendix, titled Useful Formulae.

Back to the Future

If you drove a new Shelby Mustang through a time warp to 1966, no doubt its looks alone would attract a lot of attention. But the performance would be even more likely to astonish racers of the era. A 5.4-liter engine making 500 horsepower certainly wouldn't have been impossible back then. However, such an engine would have been a temperamental beast, not something a sane person would drive on the street every day. It would have difficulty starting when cold, demand 100 octane fuel and lots of it, have completely horrible emissions by today's standards, and less than perfect behavior, except maybe at full throttle. And it certainly wouldn't come with a 36,000-mile warranty.

Electronic engine management is one of the key advances that has made it possible to turn what would have been the 1960s racing engine into a practical commuter car. By injecting a calculated amount of fuel and firing the spark plugs at precisely the right moment, today's engine control

A modified Toyota 2JZ-GTE motor can put over 600 hp to the rear wheels from 3 liters without breaking a sweat. Good tuning lets you get this sort of power without an enormous motor or building a high-strung race motor.

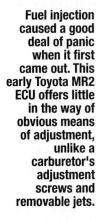

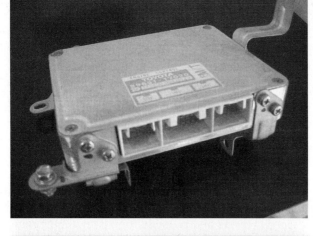

Fuel injection caused a good deal of panic when it first came out. This early Toyota MR2 ECU offers little in the way of obvious means of adjustment, unlike a carburetor's adjustment screws and removable jets.

Far from being inaccessible to the backyard mechanic, it's possible for a determined tinkerer to tune or even fabricate their own EFI. This BMW 2002 uses a mixture of homemade and junkyard parts to add fuel injection.

A look at the BMW's engine. The intake manifold is pieced together from a later BMW M10 manifold and sections from Ross Machine Racing, the ignition system came off a Ford Escort, and there's a homebuilt MegaSquirt ECU controlling the fuel and ignition.

units (ECUs) have eliminated many of the compromises from the bad old days of carburetors and mechanical distributor advances. However, when fuel injection first appeared, hot rodders feared it would spell the end of backyard tinkering. The mysterious black box didn't have any jets, springs, adjustment screws, or other traditional means of adjusting it. Where do you stick the screwdriver? Tuning the ECU would require new tools.

These tools are here now. You have several options depending on your goals. Virtually all of the options out there fall into one of three

categories. In no particular order, these are: reprogramming the factory ECU, replacing the factory ECU with a standalone aftermarket ECU, or using an add-on piggyback device to change the factory ECU's behavior without altering its inner workings.

Aftermarket ECU Tuning Choices

Some factory ECUs are tunable, either by removing a chip inside and replacing it with a new chip with a new tune loaded onto it, or uploading new tuning through the vehicle's diagnostic port. A standalone ECU is an aftermarket controller that replaces the factory ECU, or at least takes over most of its functions completely. It gives you complete programmability and usually data logging capabilities that let you record and play back the sensor readings and the ECU's internal calculations. Piggyback devices modify the engine's tune with a bit of trickery, by intercepting and changing the signals going to or coming from the factory ECU, without changing what is inside the box itself. Each of these approaches has advantages and disadvantages. Which one to pick will depend on your budget and goals, as well as how much effort you're willing to put into the car (or how much effort you're willing to buy from your mechanic). It will also depend on what kind of car you have; on some cars you will not be able to pick from all three options.

Tuning the Factory ECU—The most obvious advantage of tuning a factory ECU is that it uses mostly parts that you already have. If this is an option for your car, it can hold down costs and complexity. In many cases you can use the factory-

The tools for adjusting fuel injection: A laptop computer, wideband air/fuel meter, and communications cable.

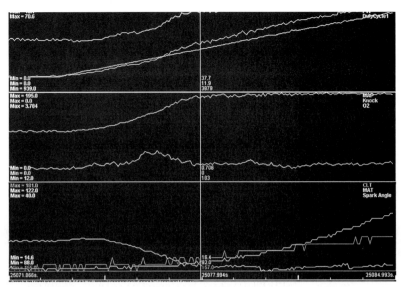

Data logging is a handy feature to have. You can record and play back sensor readings from dyno pulls or races to see how your tune is working.

The more you've modified the car, the less the factory tuning is going to fit. Saturn's engineers did not program the SL2's ECU with turbos in mind.

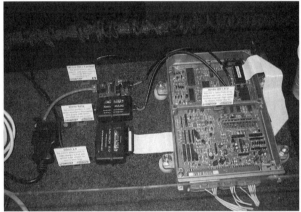

Some factory ECUs lend themselves to hacks. This Honda ECU has several extra circuits added to it from Moates and PointStep, allowing it to be tuned in real time and even control add on gauges from the ECU.

developed maps as a starting point. And in most cases all of your factory accessories will work as normal. The downside depends on what sort of method you are using to hack the original ECU.

One way to reprogram the stock ECU is to buy a prepackaged tune. It comes as a chip that you install in the ECU or a device that loads a single program onto it. These are often cheap and very straightforward to install. The downside is that these give you almost no tuning flexibility. If the tuner calibrated the product for an engine with exactly the same modifications as yours (same exact exhaust, intake, cam, etc.), this is not a problem. Some companies even sell the tuning device and a set of modifications as a package. But if you can't buy a prepackaged tune from an engine with a nearly identical set of mods to yours, the chip you buy may not be any better than the stock ECU. Also, a mail-order tune will almost certainly be a bit

on the cautious side. A tuner who has your car there can experiment and find just what settings make the best power without taking unnecessary risks, while somebody who has to work with an engine "sort of like" yours needs to be a bit more conservative to avoid breaking customers' engines that don't match the tuner's engine completely.

Port Tuners—The second option that has become more popular in recent years for newer vehicles is the ODBII port tuner that has a handful of tuned maps for different sets of modifications and fuel grades. This is sort of like having a box full of chips to choose from and some guidance to help you find one that's closest to your needs.

While it gives you more options to choose from, it also has similar drawbacks to the chip option. The map builder didn't build a map for your specific set of modifications. They may have an option for "upgraded exhaust and intake with wild cam," but any engine with a decent aftermarket following will have quite a few options in all three

Not all factory ECUs can be tweaked. At the time of this writing, there are no commercial options for tuning the stock ECU in an early Miata. Mac Spikes opted for a MegaSquirt standalone system for his Improved Touring racer to tune it for more power than the stock ECU allowed.

Most standalone ECUs were designed for racing, and come with features appropriate for a race car. Photo courtesy Big Stuff 3.

categories. And the different parts will have different tuning needs. Did the "wild" cam use the same pattern on the intake and exhaust, or go with a different lobe shape for each? Was that exhaust built for a louder, more aggressive tone, or did a real engineer design the piece with all-out performance in mind? Are the mods aimed at low-end torque, or at breathing better at high rpm? The tune you need will depend on the little details in these answers. And a canned tune, while you may see some gains, will almost never be optimum on a modified vehicle.

The third category of stock ECU hacks includes a range of devices, but they all accomplish the same thing: They allow you to rewrite the tuning information inside the stock ECU however you like. Some are ODBII port programmers with a higher lever of capability, while others are ECU hacks by companies that replace the memory in the ECU with a new chip that you can reprogram with a newly added connector. These devices often come with a steep learning curve, but they give you a good amount of tuning flexibility. There are limits to the changes you can make. An ECU may have a maximum allowable injector size, for example, and there are certain rules you can't rewrite. You're also usually limited to the inputs and outputs that the ECU already has available. Exactly where these limits are depends on your ECU and the products

available for it. In general, the pros are that you generally can gain full control of your fuel and ignition tables for a proper tune specific to your engine and modifications, you don't have to rewire your car to do it, and your accessories will continue to work as normal. The cons are the sometimes arcane interfaces available to tune some of these systems (though some of the more commercialized systems can be fairly nice). The device may only access some of the ECU's tables while leaving the more complicated tables as they were from the factory. Of the retuning the factory ECU options, this is the best of the bunch, and will meet the needs of many a weekend warrior with mild builds on a factory EFI car, if such a thing is available. But for the hardcore that needs more, a standalone is the only way to have complete control.

There is one other disadvantage that all of the factory reprogramming options can have: Reprogramming the factory ECU is only an option on certain cars. Most car companies make it difficult to access the ECU's calibration, since they don't particularly like the idea that some street racer might hack the system to gain more power, blow the motor, then erase the changes and bring it in for warranty work. The more popular the car and the fewer security measures the factory takes, the more likely you are to find a tuning solution for the stock ECU. You can't swing a spark plug wire in some tuning shops without hitting a half dozen tuning devices for LS1 Camaros, but some more obscure or complicated cars may not have any options out there. For example, at the time of this writing, there are no commercial options for reprogramming the factory ECU on a '90–'93 Miata, even though this

You can replace the factory ECU with a standalone racing ECU to get more tuneability, such as this one from the Australian company of EMS.

Sequential Motorsport ECU
EMS
8860
www.fuel-injection.com

Standalone ECUs also are a good choice for a car that never had a factory ECU. While '57 Chevies could be ordered with fuel injection, the system they used was mechanical, not electronic. Photo courtesy Edelbrock.

A look under the hood of the same '57 Chevy. Photo courtesy Edelbrock.

car has a fairly large enthusiast following.

Standalone ECUs—A standalone system can control an engine on its own, without any factory computer. These systems give you complete control over your fuel and ignition and all other aspects of engine control. Sometimes, they can take a bit more research and effort to set up because complete control also gives you more ways to set up your engine wrong. A few newer engines (where the OEM has everything including the windshield wipers, alarm system, and stereo tied into the factory ECU) may require the factory ECU to still be present in a reduced role, controlling things other than the fuel and ignition. Aftermarket ECUs are generally designed to control the engine and its related components, not some of the other wacky non-engine related bits that OEMs seem to feel the need to control with a single box or network of computers. Divorcing engine control from the factory ECU in these cases is fairly simple and the standalone can do the fun stuff, leaving its red-headed stepchild the stocker to handle the menial tasks.

Piggyback devices are another tuning option. This box from AEM modifies the signals from the ECU to the injectors and the ignition.

There are two main drawbacks to standalone systems—cost and complexity. Sometimes you can find a system designed to plug right into your factory wiring and start the car right up, but if this is not available, you can find yourself redesigning half your car's wiring and teaching the computer how to start the engine on your own. Once you do get a standalone system running, however, you'll have the freedom to tune it however you see fit. This is ideal if you're looking for all out power on your race car and want to be able to adjust everything on a dyno, or if your engine has been modified to the point where the stock ECU just plain runs out of its range of adjustment.

Being designed for racing, standalone EFI systems often come with features designed more for getting the most out of a race car than a stock ECU designed for drivability and passing emissions tests. It's not uncommon to find specialized rev limiters to make drag strip launches easy, built-in nitrous control, overboost protection, and the like in an aftermarket system.

Standalone systems are also an obvious choice for a car or engine that never had fuel injection from the factory. If you have to design your fuel injection system from scratch and calibrate it yourself anyway, an aftermarket standalone system will make the process a good bit more user-friendly. While it's sometimes possible to adapt a factory ECU off of a different engine to some other previously carbureted motor, the learning curve can be fairly steep to take on tricking a non-programmable ECU into running properly on an engine it was never intended for. In this situation standalone systems can make it easier to piece together parts from unrelated sources for your carburetorectomy, instead of requiring you to get all your hardware from one particular donor engine.

Piggyback Systems—Last is the piggyback system. This is a device that modifies the signals going to or coming from the stock ECU. Some

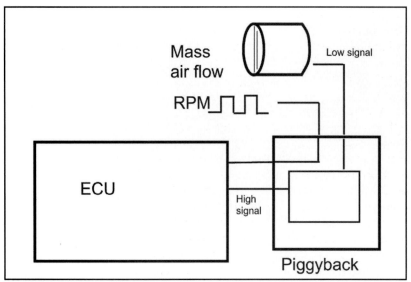

A piggyback device sits between the factory ECU and the devices under the hood, and modifies the signals going between them.

EFI system in a box. This system includes all the components of electronic fuel injection—ECU, injectors, fuel pumps, sensors, wiring, and even the intake manifold—in a single turnkey package. Photo courtesy Edelbrock.

piggyback devices fill only one function, such as disabling a speed limiter. A serious piggyback device that modifies the signals from the ECU to the ignition and injectors can give you more control than a prepackaged chip, but usually less control than either a standalone system or a device that lets you edit the factory ECU's settings at will. Devices that only modify the signals into the ECU often have a quite limited tuning capability. Their main advantages are that they are less expensive than most standalone systems (although there is a bit of overlap) and that they start with the factory settings, which can make tuning a bit easier.

Parts of a Fuel Injection System

While there are many different ways to tune fuel injection, you'll work with many of the same basic parts.

The obvious one we'd mention here is the ECU. This can also be known by other names, such as engine control module (ECM), powertrain control module (PCM), computer (C), brain box (BB), black box (BB2), stupid computer thingy (SCT), or whatever you like to call it (WYLTCI). This is the computer that calculates how much fuel to inject. In most cases, it will usually also calculate when to fire the spark plugs, and sometimes calculate a precise timing of when to inject the fuel, in addition to determining how much gets injected. Some ECUs may control other parts of the car such as timing the shifts in an automatic transmission.

In order to carry out its calculations, the ECU uses a network of sensors scattered throughout the engine compartment. These are the computer's eyes and ears. These sensors measure how fast the engine is turning, how much air is coming into the engine, the temperature of that air, and sometimes even the weather conditions. There are also usually sensors located in the exhaust that the ECU can use to check to see if it has miscalculated how much fuel to inject and adjust its calculations if necessary.

If the sensors are the computer's eyes and ears, the computer also needs hands in the form of the fuel system and the ignition system. These are the mechanical parts that give the computer the ability to use its calculations to actually control the engine.

Injectors—The main working piece of the fuel system is the injectors. They come in different shapes, sizes, and locations, but they all work in pretty much the same way. They are electronically controlled valves that can snap open or closed instantly. The ECU controls how much fuel to inject by changing how long the injectors open and close. A system of pumps, fuel lines, pressure regulators, and other devices ensure the injectors have a steady supply of fuel to feed the engine.

The ECU can also control the ignition system. The earliest such ignition systems used a single coil controlled by the ECU to generate the spark, and a rotary switch called a distributor to route the spark from the coil to the cylinders. However, with a smart enough ECU and the right sensors, the ECU can use several coils and command the coils to fire directly without a distributor. Such distributorless ignition systems have two advantages. One, they avoid wear and tear on distributors, which are the fastest wearing parts in a modern ignition. Two, they give the coils more time to charge, since the

The ECU can control the ignition, too. This Miata has four coils triggered by the ECU, eliminating the need for a conventional distributor.

coil does not have to fire as many cylinders. Distributorless ignition systems come in two varieties that each work a bit differently. The first is known as coil-on-plug (COP) or coil-near-plug, and the second is wasted spark. Both are distributor-less systems, both use multiple coils, but the number of coils is different. More on these later.

If you're looking to tune an engine that was already fuel-injected from the factory, you'll need to take the fuel and ignition system you have into consideration. Standalone system buyers will need to see what will work with an aftermarket ECU and what you'll need to swap out. If you're converting a carbureted engine to EFI, you can easily find yourself setting up all these systems from scratch. Some aftermarket EFI systems require you to use their specific components without much flexibility, others will work with just about any sensor or trigger wheel you can dig up. Finding components cheap though configuration can take a bit more work. These next chapters will help you to know just what you need to be able to fit an EFI system to your engine, properly wire it up, properly tune the system, and get the most out of it. It's going to be a fun ride...

Chapter 2
Fuel Delivery

You can bolt a throttle body in place of a carburetor, but there's more to a complete EFI conversion than just that. Converting from carbs to EFI requires changing many parts of the fuel system used with a carb.

We'll start by taking a look at the mechanical side of EFI, starting with the operation of the fuel system and related components and some of the differences from their carbureted counterparts. We'll also cover some of the basics of intake manifold and throttle body choices when running EFI.

As you probably suspected, there's more to installing fuel injection than bolting a throttle body with some injectors where a carburetor once sat. A well-designed fuel system will have every component from the gas tank to the injectors chosen to ensure the engine will have the proper amount of fuel available at all times. If you're installing a standalone ECU on a car already designed for fuel injection, many of the original parts are likely to be adequate unless you've also modified the engine to the point where the original fuel system can't supply enough fuel to keep up. If you're installing fuel injection on a car that originally had a carburetor, or need to build the fuel system from scratch anyway, you'll need to examine each part of the fuel system to make sure it's up to the task.

Fuel Tanks

The basic idea of a fuel tank doesn't sound very difficult; it just has to hold gasoline where the fuel pump can collect it. The problem is that fuel injection does not appreciate having its fuel supply cut off, even for a second. Carburetors are a bit more forgiving because their float bowls hold a small reserve of fuel. A fuel tank that worked reasonably well when it supplied a carburetor may prove to have trouble when asked to supply a fuel injection system. Typically this is most likely to rear its head under higher-*g* situations such as heavy

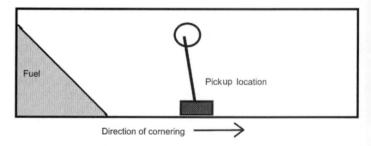

A fuel tank from a carbureted motor can have trouble if the car experiences large cornering forces, as the fuel can slosh from side to side and keep fuel out of the pickup.

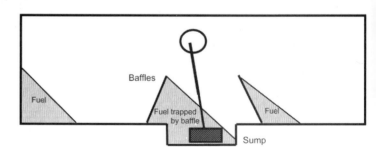

Adding baffles and a sump can reduce the fuel slosh problem.

acceleration or cornering. Fuel tanks nearly always have a single pickup tube where the pump gets its fuel. When the tank is below a certain point, halfway maybe, the fuel can slosh around under acceleration or cornering, leaving the pickup high and dry with no fuel to draw into the lines. Having the fuel slosh forward under braking is less of a concern, since the engine will usually have the throttle closed or may even have the ECU shutting off the fuel to the injectors under braking, though some cars may still have issues if you are not able to get any fuel under braking. The pickup will still need to be someplace where it can get fuel when you're parked on a downhill slope trying to fire that engine up, too.

Using a Sump—If it's practical to design a gas tank for EFI from the start, or make extensive changes to an existing gas tank, the usual solution is to use baffles and possibly a sump. A sump is a deep area on the underside of the tank to hold the pickup tube. The sump is usually located at the center rear of the tank, although circle track cars usually have the sump offset to the right because they only have to turn left. Baffles are pieces of sheet metal that make it easier for fuel to splash into the pickup tube area than out of it. A car built entirely for drag racing may be able to get away with a shallow, rear-mounted sump, or even run a tank never designed for EFI, while corner carvers often need more baffles.

Surge Tanks—Sometimes it isn't practical to modify the gas tank and you have to instead design a fuel system so that it continues to supply fuel to the engine even if the pickup sucks up air. If you'd rather not replace your tank with a suitable tank or fuel cell, you have another option called a surge tank. A surge tank can separate out the air and ensure a smooth fuel supply delivery even if air is being sucked up from the main fuel tank on occasion. The surge tank is a sort of extra fuel tank, usually a tall and narrow design. A low-pressure fuel pump feeds fuel into the top of the surge tank, while a high-pressure fuel pump picks up fuel from the bottom of the tank and supplies it to the engine. A third line on the surge tank, mounted at the very top, feeds back to the main fuel tank. The low-pressure fuel pump supplies more fuel than the high pressure pump, so when the engine is running normally, the excess fuel circulates through the return line. If the supply to the low-pressure pump is cut off for a short time, the high pressure pump can continue supplying fuel until the surge tank empties. If all goes well, the low pressure pump will start sending fuel before the tank is empty, and all is well. While a surge tank may seem less difficult than

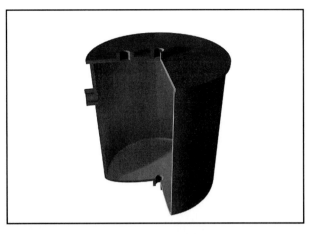

A surge tank is another way to deal with fuel slosh. Making it tall and narrow can prevent the fuel from sloshing very far.

removing the main gas tank and having a sump and baffles welded in, surge tanks require more complicated plumbing and an extra fuel pump.

Fuel Cells—Purpose-built race cars often use a fuel cell. That term has become a bit of a buzzword of late, but this isn't the sort of fuel cell that turns hydrogen into electricity; instead, a gasoline fuel cell is a type of fuel tank that holds a spongy foam inside a rubber bladder. It is less likely to rip open in an accident, and if it does, the fuel leaks out more slowly than from a conventional fuel tank. Some racing classes mandate using fuel cells, so check the rulebook to see if you can use your production fuel tank or will need a cell. Although it's often possible to route a hose from the stock

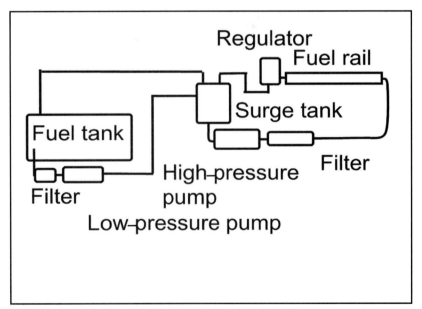

Plumbing for a surge tank. The surge tank gets its fuel from a low-pressure pump, while a high-pressure pump located below the surge tank carries the fuel to the engine.

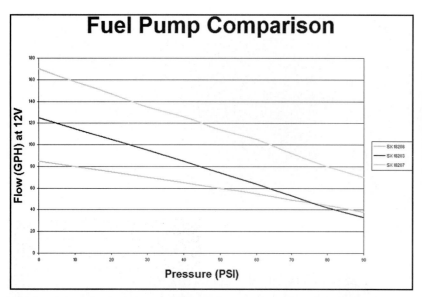

Fuel pumps deliver less fuel as pressure increases, and more fuel as voltage increases. This graph shows the fuel flow for SX Performance's line of EFI fuel pumps. Note that the smallest pump actually keeps up with the mid-sized one at high pressure, while the midsized pump does better for large naturally aspirated motors.

Aerometive makes a complete sump kit that can be welded into a new gas tank.

filler door to the neck on a fuel cell, the fact that commonly, a fuel cell is mounted higher in the chassis than the original tank, can make it difficult or impossible to fill from the stock filler door. Most often, you'll fill the fuel cell by popping your trunk to access it directly. Also, if you do use a filler neck your class rulebook may require replacing the filler neck with a dry-break connection. Plan your installation carefully.

Fuel Pumps

The fuel pump needs to be up to the task of feeding the demands of the engine. It needs to supply enough volume for the engine and at adequate pressure for your injectors. A fuel pump's flow capacity drops off as the pressure climbs, so a pump advertised as flowing 80 gallons per hour at 43.5 psi may not flow nearly that much when asked

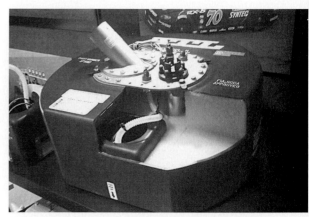

Cutaway view of a fuel cell shows the foam inside.

to supply the fuel at 65 psi to a turbo motor.

Sizing a Pump—The fuel pump needs to supply more than enough fuel to feed the engine at full power, since the fuel system returns part of the fuel to the tank all the time to regulate the pressure. However, the fuel pump should not move too much fuel, or it will heat up the fuel, an undesirable side effect that can evaporate some octane-boosting chemicals. Sometimes manufacturers make it easy by rating their pump for a suggested horsepower application, but often you just have the flow ratings to go by, generally with different ratings for naturally aspirated and forced induction use. Since most fuel pumps flow less fuel at high pressure, a pump that's just adequate for a 300 hp naturally aspirated V-8 may not be able to feed a 300 hp turbocharged four-banger even if they each need the same amount of fuel. This is due to the turbo engine, while at high boost, running the pump at higher pressure where it may not be able to provide the same flow rate. More on this in a minute.

A common tuner rule of thumb is to make the pump large enough to supply 130% of the fuel the engine will burn at full throttle. You'll find further formulae on sizing a fuel pump on page 152. We don't want to bog you down with that here but it's there when you need it.

Fuel injection runs at a higher pressure than a carburetor. Throttle body injectors typically run at 10–12 psi, while port fuel injectors typically run at about 43.5 psi. Keep in mind that the pressure that matters to the fuel injectors is not the absolute pressure, but the difference in pressure between the fuel in the fuel line and the air inside the intake at the injector nozzle. That means in a boosted application when you add boost, you need to add fuel pressure to compensate for that boost, so that the fuel continues to flow at the same rate. So for every pound of boost, you'll need an extra pound of fuel pressure, and this pressure difference is quite

This aftermarket external fuel pump can feed an engine making up to 1,500 hp.

Many OEM fuel pumps fit inside the fuel tank and use the fuel for cooling. This test bench keeps the fuel pump submerged in a transparent tube.

Mechanical fuel pumps can move large volumes of fuel without the electrical current draw. This Racepumps unit runs off the stock mechanical fuel pump drive.

This belt-driven fuel pump is another approach to using mechanical fuel pumps with EFI.

important to the pump, since the fuel in the tank is at atmospheric pressure. More on this in a moment when we discuss fuel pressure regulators further.

Types of Fuel Pumps—Most factory fuel pumps on cars with factory fuel injection are located inside the fuel tank, which can help cool the pump because the gasoline carries the heat away. The downside is that an in-tank pump is sometimes a royal pain to replace unless the designer took the time to provide an easy way of reaching the pump. In-tank pumps are often quieter than external pumps, although with careful mounting and sound insulation, you can make an external pump quiet, as well. Many late-model Mercedes demonstrate this quite well. If you are using an external electric fuel pump, it's best to mount it below or near the bottom of the fuel tank so that the inlet is fed by gravity. Most electric fuel pumps are good at pushing the fuel, but not very good at pulling it into the pump. Some pumps are better at drawing fuel than others, and in most cases if you keep it below the top of the fuel tank you'll be fine. Some cheap pumps though absolutely will not self-prime unless they are gravity fed. Best to mount those types as close to the bottom of the fuel tank as practical.

Most fuel pumps for EFI are electric. Electric pumps provide fuel pressure as soon as you turn the key, and they're made in large enough quantities to bring the price down. Companies such as SX Performance, Aeromotive and Fuelab manufacture electric pumps designed to support up to 1,500–1,800 hp. However, an electric pump big enough to feed a 1,500–1,800 hp engine will draw some serious current, putting a strain on some vehicle's electrical systems (not to mention requiring somewhat heavy wiring). In some cases

you may want to look into using a mechanical fuel pump designed for EFI. These pumps are usually belt driven like an extra alternator, but some designs such as a pretty interesting newer design from RacePump use the same mounting pad and drive mechanism as a factory carburetor fuel pump.

The biggest drawback of mechanical fuel pumps is that they don't deliver fuel pressure when the system comes on. It can therefore take a good bit of cranking to build fuel pressure and get the car to light off, so in some cases starting a car with a mechanical fuel pump can be difficult. This isn't always a big deal on a pure race car, a bit of starting fluid and you're on your way, but that's not what EFI is all about—for the best of both worlds, you can go so far as to use an electric pump to pressurize the fuel system instantly for quick starts, and a large mechanical pump for the heavy work. A one-way check valve can be used in the fuel system. This allows the electric pump to pressurize the lines (generally via a switch) but not allow fuel from the mechanical pump to backflow into the electrical pump once the engine is started and the electric pump is deactivated.

A tubing bender puts precise bends in metal fuel lines.

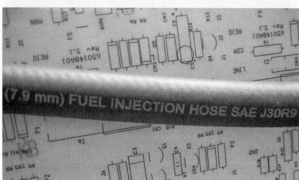

If you're using rubber fuel hose, be sure it's rated for fuel injection. If it isn't marked "Fuel Injection Hose," assume it isn't.

Fuel Lines

There are many options for fuel lines. Which one to use will depend on your budget and what it says in the racing rulebook. Fuel lines come in two basic types: hard lines that stay in the shape you bend it to, and flexible lines. While it is possible in some cases to plumb a car with only flexible hose, plumbing a car entirely with hard line is not a good idea. You need to build some flexibility into the system to handle vibration. Even purpose-built race cars with solid motor mounts usually have some flexible line in the system.

You should keep in mind that some materials that look like they can be used as fuel lines do not belong on a fuel-injected car. Hoses designed for carburetors are not rated for the pressure an EFI system will generate in the lines. And copper tubing acts as a catalyst on gasoline. Contrary to what some hucksters will have you believe, metal fuel catalysts aren't a good thing: Copper promotes having the gasoline turn into gum and varnish.

Hard Fuel Lines—Hard fuel lines typically come in three types: aluminum, stainless steel, and mild steel. Aluminum line is soft and bends easily. It's also very lightweight. You will need to support it at regular intervals along its length as long unsupported lengths of aluminum will fatigue and crack when exposed to the vibrations in a car. Its other

downside is that it does not handle corrosive fuels like methanol or E85 very well. Stainless steel is very strong and corrosion resistant, but expensive and a bit harder to work with. Use this if you expect to run alcohol, since alcohol will absorb water from the air and this can corrode other types of fuel lines. Mild steel is a reasonable option for a budget; many factory fuel lines are made from this.

Flexible Fuel Lines—There are several types of flexible fuel line, too. Rubber hoses are quite common, but you will need to make sure the fuel hose you buy is appropriate for the pressure you find in fuel injection. If you are allowed to use rubber hose, you'll want hose rated at SAE 30R9 or better. The less expensive SAE 30R7 category of fuel line is a low-pressure sort of hose meant for work with carburetors. However, many racing sanctioning bodies view rubber hose as a safety hazard because it is easy for a bit of sheet metal to cut the fuel lines in an accident. For example, NHRA safety rules do not allow more than a total of 12" of rubber fuel line in the entire fuel system. And they do not permit it in certain especially hazardous locations, such as near the clutch (to prevent a flywheel explosion from also starting a fire). Also, you should never, ever run the fuel lines inside the passenger compartment. Keep the fuel line where a leak won't spill on you.

Safety rules generally require braided hose instead of rubber hose. The most recognizable version of this hose has an outer stainless steel layer over a rubber inner layer. Some versions of braided hose add an extra layer of rubber over the stainless steel braid, since the braided steel can saw through other parts over a lifetime of vibration and movement. Other versions replace the stainless steel with Kevlar and Nomex.

As a general rule, 3/8" fuel lines are big enough for up to 500 hp, 1/2" fuel lines can feed up to 800 hp, and 5/8" lines are good for up to 1,100 hp. These apply for gasoline; fuels with a lower air/fuel ratio requirement, like methanol, require larger lines. If you are using AN hoses, the number after the dash equals its size in sixteenths of an inch. A –6 hose, for example, is 6/16" or 3/8" inner diameter.

You'll find a variety of fittings to connect hard lines to hoses and to other parts of the fuel system. Flare fittings can let you connect a hard line to an AN hose fitting, but making a leak free flare connection takes practice. Compression fittings can be a bit easier to assemble but are sometimes more expensive. While compression fittings are not safe for the high pressure found in brake lines, good quality compression fittings are safe for fuel lines.

How to Assemble a Hose End

While sanctioning bodies do not necessarily require using AN fittings with braided hose, they're what most installations use. It's also possible to attach braided hose with hose clamps, although the braids tend to fray out. Another, lesser-known standard for hose ends is Joint Industrial Council (JIC). JIC fittings are usually used on industrial hydraulics. It's possible to connect AN and JIC fittings as their specifications for flares and threading are the same. Be warned that the 45 degree brake flare fittings are not the same as AN flares at all. These fittings sometimes physically will bolt together, but they will leak if you do this.

Attaching a hose end starts with cutting the hose to the right length. The braided steel layer around the hose can make this step tricky. Cut it wrong, and the braid becomes tattered and you will have a very hard time getting the braid into the socket. If you don't have a hose cutter specifically made for braided hose—which might be a good thing to buy if you're using these types of hoses often—the most effective way to cut the hose is to wrap it in a layer of tape and cut through it with an abrasive cut-off wheel. Once it's cut, make sure to clean up any dust from the inside of the hose.

Most AN fittings come in two pieces, a socket and a nipple. The socket goes on first. You'll need to put the hose in a vice and twist the socket on, as it has threads inside that grab the outside of the hose. Then grab the socket with a vice and poke the nipple through the socket and thread it into the inside of the hose. Tighten it carefully with a wrench and you're finished. Once you've got the fitting onto the end of the hose, it simply threads onto other AN fittings.

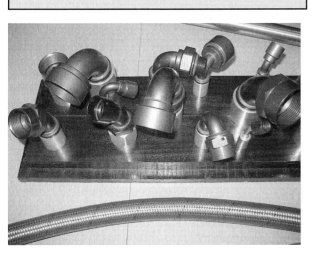

A spin-on canister-type fuel filter is quite similar to an oil filter in appearance.

An inline filter made for racing applications.

Fuel Filters

Dirt or other junk in the fuel can easily clog a fuel injector. Fuel filters are rated in microns, which measure the size of the smallest particle they can catch. The smaller the number, the finer the filtration. The fuel filter needs to have both a flow rating and a pressure rating high enough to keep up with your fuel system. If you're fuel pump is inside you're fuel tank, you're typically have a single fuel filter downstream of the fuel pump, meaning outside of the tank along the lines somewhere. If you're converting an older carbureted vehicle to EFI and using an inline fuel pump it's a good idea to use a pre-pump filter in the range of 100 microns and a post-pump filter that's much finer, about 30 microns or smaller.

Filters for EFI come in several different shapes and sizes. Some designed for racing use have AN fittings built into the ends, while factory filters usually have hose barbs for rubber hoses. Some industrial filters have pipe threads instead. As a general rule, the larger filters can be expected to last longer, while smaller ones are typically more race oriented items that need changing more often. Some canister type filters are large enough they can even double as a surge tank if you mount them with a tube that goes to the bottom of the filter as a pickup for the high pressure pump.

Fuel Rails

Most fuel injection systems use a fuel rail to deliver the fuel to the injectors, or two fuel rails for V-type engines. The rail is simply a large diameter tube with holes in its side for delivering fuel to the injectors. The fuel enters the rail at one end, and leaves the rail at the opposite end through the fuel pressure regulator. In the case of a typical V-type

A pair of fuel rails for a V-8.

A stock fuel pressure regulator: Not adjustable, but in this case it can still keep up with the rest of the engine build.

engine with dual rails, fuel will enter at the end of one rail. The other end of that rail will be connected to the second rail via a fuel line, and the regulator will be at the far end of the second rail regulating pressure for the entire assembly.

Occasionally, you'll see fuel rail layouts where the fuel line splits before the rails and each rail has its own inlet and outlet. The inner diameter of the fuel rail needs to be at least as large as the inner diameter of your fuel lines, and making it a bit larger won't hurt.

Some setups do not use a rail in the conventional sense. It's also possible to use a distribution block, with hoses or hard line running from the block to the injectors. This block is similar to a fuel rail in that it also has the fuel delivery on one side and the pressure regulator on the other. Another setup you might run into are TBI units with built-in fuel pressure regulators such as GM's TBI units used through much of the '80s and early '90s. In this case there is no external fuel rail, but the internal passages of the TBI unit act as the fuel rail and the unit has a pressure regulator built right into the throttle body unit.

Fuel Pulsation Dampers

Sometimes you will see a device in the fuel system called *fuel pulsation damper*. This does exactly what its name implies, helping to reduce pulses in the fuel supply system. It can also sometimes be called an accumulator. These use either an elastic material or a trapped air bubble, which lets more fuel into the damper when fuel pressure spikes and releases fuel when the pressure drops, helping to smooth out the fuel pressure. The pulsation damper can be anywhere between the fuel pump and the pressure

regulator. As the on/off nature of fuel injectors causes a lot of the pulsation, you'll often find these attached to the fuel rail. Not all EFI systems use these, and some fuel pressure regulators have the pulsation damper built in.

Fuel Pressure Regulators

While carburetors often use a "dead-head" regulator upstream of the carburetor limiting the pressure downstream, EFI normally uses a regulator that limits the pressure upstream. The fuel pressure regulator (FPR) has a valve that opens as needed to keep the pressure in the fuel rail at a set level by sending excess fuel back to the gas tank through a return line. A regulator that's too small for the fuel pump will cause the pressure to rise too far, particularly at idle when less fuel is going through the injectors.

A normal regulator needs to maintain a constant pressure difference between the fuel rail and the intake manifold, not a constant absolute pressure. This is particularly true on forced induction engines, where the boost pressure would make it harder for the fuel to come out of the injectors as the boost goes up. The regulator has a vacuum port for this purpose. This port should be connected to a vacuum line downstream of the throttle body, unless you have injectors that mount upstream of the throttle blades, where the regulator needs to see the pressure from the same area where the injectors are located.

Most fuel pressure regulators, particularly those intended for forced induction applications, but also many used in OEM non-forced induction applications, will handle this task for you with no problem. The regulator will reference intake manifold vacuum/boost. Under vacuum, the regulator will lower fuel pressure, while under boost the regulator will increase fuel pressure in a 1:1 ratio. For each pound of boost, the regulator will add a corresponding pound of fuel pressure. This way, if you've set your base fuel pressure to 43.5 psi, for example, and you've set the boost controller to 20 psi, then when you're running at wide-open throttle (WOT) and 20 psi of boost in the intake your injectors are seeing about 63.5 psi of fuel pressure. This is a 43.5 psi difference over the air pressure in the intake, allowing the fuel to flow similarly to how it would at 43.5 psi fuel pressure with no boost in the intake. Note there are also adjustable and high rise fuel pressure regulators that raise pressure at much higher rates than 1:1. These are typically used as an attempted fueling solution for forced induction applications without the luxury of control of a tunable EFI system. They are

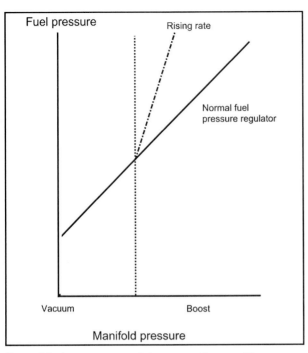

Fuel pressure

Rising rate

Normal fuel
pressure regulator

Vacuum

Boost

Manifold pressure

Normal fuel pressure regulators are referenced to manifold vacuum and add one psi of fuel pressure for every psi of vacuum. Rising rate fuel pressure regulators add more fuel pressure when in boost.

This Magnafuel regulator does not bolt directly to the stock fuel rail, so it uses an adapter and hose to connect.

known as *rising rate* fuel pressure regulators. A rising rate fuel pressure regulator would add more pressure than a normal regulator, such as 6 psi of fuel pressure for every 1 psi of boost. These normally show up on engines with a stock ECU that is unable to compensate for boost. Some people choose to run rising rate FPRs with a standalone EMS in an effort to force more fuel through injectors that are too small to handle the load at wide open throttle.

In most cases I'd suggest you just replace your injectors with adequately sized units. There is a special case though: Let's imagine for a moment you're setting up a motor with large fuel needs under high load that would normally require very large injectors. That could affect your idle quality, but you want it to purr like a kitten at idle, so that's not an option. Well, in some cases you have no choice but to go to a staged injection setup, but that's another topic. In milder scenarios you could run slightly smaller injectors and a rising rate fuel pressure regulator that allows for a smoother idle, but with the additional fuel pressure will flow what you need under heavy load. If you're using a rising rate fuel pressure regulator, keep in mind that you'll be running a higher fuel pressure and size your fuel pump to match. Keep in mind what we mentioned earlier: At higher pressures, your 50 gph pump will not flow 50 gph anymore, and you'll have to size

appropriately for your application. Most fuel pump manufacturers can provide a pressure vs. flow curve for their pumps.

If you're using an adjustable fuel pressure regulator, you should set the pressure with the engine off and the fuel pump running. Alternately, you can set the pressure at idle with the vacuum port on the FPR disconnected so that it's not referencing vacuum and lowering the fuel pressure based on this reading. You'd set your base fuel pressure, typically to about 43.5 psi as a starting point, and then hook the vacuum port back up.

Returnless Fuel Systems

One recent trend in fuel system design from OEMs is the returnless fuel system. These can be hard for a backyard builder to design from scratch, as without very careful design you can have severe vapor lock problems. When the engine shuts down, the fuel in the rail can overheat and boil if the fuel is not isolated from the heat, and without a return line the vapor has no place to go. If you buy a car with such a system, note that there are a couple different ways manufacturers have gone about building these systems, and this can affect what you need to do when changing your engine management.

The most common type uses a pump and regulator that both mount at the fuel tank. From a wiring standpoint, they're not much different from a conventional fuel system. You just need to switch the pump on. Upgrading their fuel flow can be a bit of a challenge if the stock pump can't handle high flow rates, however. A rarer system uses a variable speed electric pump and eliminates the fuel pressure regulator altogether. Controlling such a system requires a fuel pressure sensor in the fuel rail and an ECU specially designed to control a variable speed pump.

A fuel injector. This one is made by Delphi and is designed to fit in place of the Bosch EV1 series injectors used on many German and American cars.

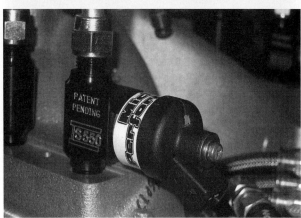

Conventional fuel injectors are available in up to 160 lb/hr flow rates. These enormous injectors from Mick's Performance flow up to 850 lb/hr.

Injectors

Now the fuel has reached the heart of the fuel injection system, the injectors. The fuel injectors are electronically controlled valves that open when you apply an electric current across their terminals to let fuel flow through the injector and into the intake manifold. There are several different injector designs out there, but all conventional injectors have an electromagnet that pulls on some sort of valve, which may be needle shaped or disc shaped. A needle shaped valve is called a *pintle*.

How Injectors Work—The fuel sprays out of the injectors at a high enough pressure to break the fuel up into tiny droplets, much like a squirt out of a cologne bottle. These tiny droplets burn much better than big, huge blobs of gasoline. A well-designed injector will get good *atomization*—that is, the drops will be as small as possible.

The fuel injectors spend most of their time either completely open or completely closed. While there is a short amount of time when they are going from one end to the other, the computer does not attempt to manage the fuel flow by holding the injector partially open. Instead, the computer controls the injectors by changing the length of time the injectors stay open, opening the injectors for a longer time to inject more fuel. The amount

of time the current runs through the injector is called the *injector pulse width*. Another useful measure is the *injector duty cycle*, which is the percentage of time the injector is open. Most injectors should not be run at more than an 85% duty cycle.

Even if your engine came with injectors, you may need to pick a replacement set. Besides checking to be sure the new injectors actually fit in the same space as the old ones, you will need to pick the injectors based on their *flow rate*, their operating pressure, and their impedance.

Injector Flow Rate—The flow rate is the first and sometimes only thing many people think about. The injectors need to flow enough fuel to produce all the power the engine is capable of. It's better to have the injectors a little to the large side, in case you need more fuel than expected, rather than to have the injectors too small and risk coming up short. However, getting carried away and overdoing the size of the injectors can also be trouble. An injector that's too large may release so much fuel at idle that the injector pulse width needed is scarcely longer than the time the injector requires to open, giving you less control over the air/fuel ratio at idle. Use a bit of common sense. A set of injectors that are 50% too big aren't likely to be a huge problem, and may even be a good move if you're planning on more power later on. But a set of 160 lb/hr injectors on a stock '95 Honda Civic is asking for trouble even if you have big plans for it later. You may find tuning injectors that much out of proportion to the engine's fuel needs will be frustrating enough to give up on those big plans.

Injector Pressure—There are limits to how far you can change the fuel pressure. In fact, it's a rare injector that you can actually run at four times the rated pressure. Most injectors designed to run at multi-port EFI pressures (around 43.5 psi) tend to have the fuel come out in unburnable dribbles at 20 psi or less. At the same time though, too much pressure can keep the valve from opening. Exactly where too much pressure is depends on the injector; some will work with 90 psi, while others jam closed at 55. There are a few low-pressure injectors that typically run at 10 to 20 psi. These usually show up in throttle body applications and puff the fuel out in a mist in all directions or a very wide cone pattern rather than squirting it out in a line or narrow cone shape like port injectors.

Remember, it's the difference in pressure between the ends of the injector that counts, not the absolute pressure. Injectors that max out at 70 psi on a naturally aspirated engine may still work properly if they are running at even 75 psi of fuel

Sizing Injectors

Picking injectors calls for working backwards from the amount of horsepower you hope to get from your engine mods. This uses a number called brake specific fuel consumption (BSFC), which is how much fuel your engine needs in a given amount of time to make a given amount of horsepower. A common unit for this is pounds per horsepower-hour. Since most injectors are sized in pounds per hour, you just need to multiply your horsepower goal by its BSFC (which is pretty similar from engine to engine) to find the total fuel flow you need. You then will need to factor in the number of injectors and their duty cycle. So your equation looks like this:

injector flow rate = (target horsepower x BSFC) ÷ (duty cycle x no. of injectors)

Typically, you can use 0.45 lbs/hp-hr for a naturally aspirated engine, and 0.55 lbs/hp-hr for supercharged or turbo motors, and 85% is a safe value for duty cycle. So if you had a V-8 with eight injectors and wanted 400 naturally aspirated horsepower, this equation would give you a minimum injector size of 26.5 lb/hr.

The metric measurement for injectors is cc per minute. One lb/hr is roughly 10.5 cc/min. The engine in the above example would need 278 cc/min injectors.

Besides knowing the flow rate, you'll need to know the injector's operating pressure where the flow rate was measured. You can run an injector at a different pressure from the rated pressure, but this changes the flow rate. Doubling the pressure does not double the flow rate; you need four times the pressure to get twice the flow.

actual injector flow = rated injector flow x

√(actual fuel pressure ÷ rated fuel pressure)

Cutaway view of a port-injected intake system. The injectors fire into the intake runners, aimed at the ports.

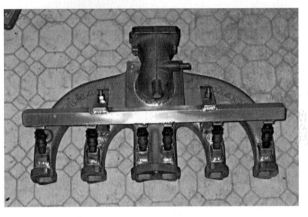

A carbureted intake converted to port fuel injection. Changes include fabricating an adapter for a throttle body, welding in bungs for injectors, and a fuel rail made of square aluminum tubing.

the people who build them, and installers just focus on the injector's resistance as measured with an ohmmeter. Usually, low-impedance injectors are 4 ohms or less, and high-impedance injectors are more than 10 ohms. High-impedance injectors are also known as saturated injectors because you saturate them with current to hold them open. Low-impedance injectors are sometimes called peak and hold injectors because these take a high peak current to pop them open followed by a lower amount of current to hold them open. Generally speaking, you'll commonly find small to mid-size injectors in high-impedance designs, up to about 50–60 lb/hr. Sizes larger than this are generally low-impedance, and there is some crossover in the middle of the range, i.e. 42 lb/hr low-impedance injectors.

Placing Injectors—There are couple of different places where you can put the injectors in the intake. Most fuel injection systems for the last decade have been port fuel injection, with the injectors located in the manifold near the ports. Usually there's one injector per port, although you can add more if you need it for the extra flow. Some early EFI systems used throttle body injection, which had one or two large injectors in a throttle body. There are also aftermarket throttle body injection systems that may use four or sometimes even more injectors in

pressure while there is 15 psi of boost in the intake manifold. While the valve has to open against fuel pressure, it has the air pressure pushing it in from the other side equalizing part of the pressure; it's the difference that matters here.

Injector Impedance—The last injector characteristic to consider is its impedance. *Impedance* is a term that combines the injector's resistance and its inductance. However, in practice, nobody measures the injector's inductance except

Using Low-Impedance Injectors on an ECU Designed for High-Impedance Injectors

Got an ECU that only works with high-impedance injectors, but the only injectors that flow enough fuel that you can find are low impedance? You might not need a new ECU. Instead, you can use injector resistors to limit the current through a low-impedance injector and let it work with an ECU meant for high-impedance injectors. The best way to wire these up is to wire one resistor in series with one injector. You can calculate the current flowing through the resistor/injector combination with Ohm's Law:

V = I x R, where V stands for voltage, I equals current and R is resistance

In this case, you'd divide the voltage (figure on 14 volts for a normal 12-volt electrical system, since the actual voltage is a little

Injector resistors are one way of dealing with low-impedance injectors. This resistor pack is from a Toyota inline six.

higher) by the total resistance of the injector and the resistor to find the current. When in doubt, make the resistance of the resistor high enough so the total resistance is about equal to the resistance of the stock injector.

You'll need a high current resistor; the tiny, spindly things you'd find inside the ECU are not up to this sort of current. So you will need to calculate the power draw in watts. Multiply the current you've calculated by the square of the resistance, and that's how much power the resistor must handle. Get a resistor rated for slightly more watts than that to be safe. Often you can find a suitable resistor in a junkyard; look for finned aluminum gadgets about the size of a pack of cigarettes.

There are two potential downsides to injector resistors: They cause the injectors to open a bit slower, and they can lower the maximum pressure the injector can handle before it stays closed.

A better option is to install a peak-and-hold driver box between the injector and the ECU. This box reads the signal from the ECU and opens the injector for the same length of time the ECU commands it to, but after a set interval, it turns on a current limiting mode to keep the injector open using less current. These little ditties let you keep your ECU and get around the potential drawbacks of injector resistors. Their main downside is their expense as compared to a few resistors.

the throttle body. Throttle body injection is fairly cheap to build and makes converting from a carburetor a bit easier, but this can have fuel distribution issues, much like a carburetor might have. Much of this is due to the wet flow design of the intake manifold flowing both fuel and air together. Hence a major benefit of multi-port injection: the manifold flows only air up until just before the port, when the fuel is added to the mix.

Injector Timing—There are several different ways the computer might control the injectors. You might expect the computer to try to time each injector to fire when the intake valve it points at is open. Many systems do implement some form of timing, although it may be timed to spray onto the closed (and hot) valve to vaporize the fuel better, often just before the valve opens. Sometimes it continues to spray as the valve opens, but stops the spray before the valve closes. Such a system is called *sequential injection* because the sequence of injector firings match the sequence of the cylinders.

A computer can also fire the injectors all at once and not time them to anything. This is called *batch firing* the injectors. You might expect this to cause problems, but carburetors have gotten away with untimed fuel delivery for over a century. The fuel eventually gets sucked down the intake valve one way or another. A variant of batch fire is *bank to bank*, where the injectors are divided into two groups that alternate firing.

Sequential firing sounds like it would be ideal, and it is. However, it does not make significantly more power in production engines than a suitable batch/bank fire system; the difference may be three to four horsepower on a 500-horsepower engine in the hands of a master dyno tuner. The reason is that at full power the injector stays open so long that it is still spraying at a less than desirable time for most of its cycle. Sequential injection also can sometimes idle with a slightly leaner mixture, which is good for emissions and may give a minor boost to fuel economy. However, a bank-to-bank system can still get very good power and drivability for less money. It depends on whether that last percent is critical or if you're trying to get the most power for your dollar.

There are pros and cons to each to be considered. Sequential is going to make that last percent possible, and give you the ability to tune for slightly better emissions and fuel economy. It's a bit more complex to install properly, and to tune. To get that last percent or two of power will require extra dyno time and instrumentation to determine which cylinders may need a bit more or less fuel to wring out every bit of power. This tuning time will not be cheap, nor will the instrumentation to give the tuner the info they need to do their job. Batch or bank to bank on the other hand will get the majority of us pretty much the same power numbers for less money and less complexity. It comes down to the target vehicle, your budget, and any need you might have to always run the latest technology you can run—is it a low buck cruiser, a maximum effort race machine, or something in between?

There is one type of engine that needs injector timing control a bit more than others. Some engines, like the ones in old Minis, have Siamese ports—two cylinders sharing a single intake port. On these engines, the fuel intended for one valve can get sucked down the other, in an effect called *charge stealing*. Such an engine either needs precisely aimed injectors and good timing control, or injectors located so far from the ports that the fuel mixes with the air enough that charge stealing is not likely to be a problem. To a smaller extent this can be an issue on other engine types, but on a Siamese port engine it will be a bigger issue. A proper sequential system setup to fire the fuel at the right time for each cylinder can avoid this problem.

Manifolds and Throttle Bodies

Carburetors require manifold designs that can be as much voodoo as science. An intake manifold for a carburetor needs to prevent the fuel from dropping out of the air/fuel mixture, keep up the speed of the incoming air through the carburetor to get a good vacuum signal, and keep one cylinder's fuel from accidentally falling into a different cylinder. Throttle body injection setups still have to deal with many of these issues, although at least they don't have to worry so much about a vacuum signal. But having the fuel injected at the ports lets the manifold designer focus on a single concern: building a manifold that will deliver the air with the least restriction in the rpm range where the engine operates.

Manifold Types—There are two common types of manifold designs used on port-injected engines: the plenum-runner design and independent throttle bodies. Both of these deliver air to the cylinders through tubes called *runners*. In theory, an engine could have the cylinder ports open to the outside air and still run as long as you found a place to put the injectors and a way to control the amount of air coming in. If the flow through an engine were completely steady, the extra length of the runner would be a restriction. But properly sized runners improve the flow on a real engine. The air moving through a tube has momentum, and when the intake valve closes the momentum can keep the air

A throttle body injection system puts the injectors at the throttle body, much like an electronic version of the carburetor.

Independent throttle bodies offer great throttle response and distinctive looks.

moving through the runner towards the closed valve so there's a high pressure charge of air waiting for the valve to open. An intake with no runners would require the air to enter in abrupt starts and stops, which would make the flow considerably less smooth and reduce power.

The runners are often tuned to take advantage of pressure waves at a particular rpm. As a general rule, longer runners make more low rpm torque. At high rpm, however, a longer runner becomes restrictive and hurts power. The Chevy TPI manifold is a prime example of a long runner manifold that gives a large hit of torque at low rpm, but starts to lose power at about 5,000 rpm.

Plenum Runner Manifolds—A plenum-runner design has the runners connected to a large open chamber, which is the plenum. A single valve called a throttle body lets air into the plenum, and tubes branch off the plenum to deliver air to the cylinders. If you're familiar with the manifolds for a Holley four-barrel carburetor, this idea is very

EFI for Carburetor Manifolds

If you're starting with an older engine design, you may find yourself having to rework an intake originally designed to use a carburetor. Some less frequently modified engines do not have any off the shelf intakes built for injectors. Even some engines that have factory EFI intakes have a better selection of carbureted manifolds when it comes to airflow.

The simplest way to carry out an EFI conversion is a route that the manufacturers themselves tried when they first added fuel injection: Replace the carburetor with a throttle body injection system. As a throttle body injection system is self-contained, it requires minimal changes to install. If you have a popular carburetor type, like a Holley four barrel or Weber DCOE, you can find several designs that will bolt right to your manifold. More unusual carburetor types like a Carter BBD will usually require some sort of adapter plate to get the bolts to line up, or possibly just drilling the intake manifold for new studs.

Throttle body injection can give you a lot more control over the tuning than a carburetor, and it's almost always cheaper than a multipoint fuel injection setup. If you had relatively good fuel distribution with a carburetor, a throttle body injection unit may be just right for a budget buildup. However, if you already have fuel distribution issues with your carburetor, they could just as easily get worse when you put a throttle body injection system on there. If you already have uneven fuel distribution you're more likely to benefit from port fuel injection. If your distribution problems prove to be bad enough, or you otherwise have the need to take advantage of individual cylinder tuning, you'll want to step up to sequential injector control, on top of your port fuel injection intake and injectors.

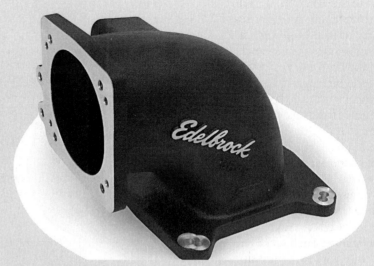

Elbow adapters let you install a throttle body on a carbureted manifold. Photo courtesy Edelbrock.

Another option for putting a throttle body onto a carbureted manifold is to use a throttle body with a carburetor-like bolt pattern, such as this one from BBK.

A port fuel injection conversion generally takes a bit more work than throttle body injection. If you are starting with a modern manifold design, it may already have bosses cast in for fuel injectors that just need to be drilled and reamed to fit your injectors. If these are not there, you'll often need to drill oversized holes and weld, thread, or epoxy in a set of injector bungs—ready-made pockets to accept your fuel injectors. It's generally best to aim the injectors at the intake valve, but sometimes the shape of the manifold requires you to point them at a shallower angle or even perpendicular to the runner. Once you have the injectors in place, the next step is to fabricate a fuel rail and hold down clamps to keep the rail in place. Several aftermarket companies supply extruded stock just for this purpose, or you can use square aluminum or stainless steel tubing with bungs welded in place. Some injector bungs let you skip the rail and simply run a flexible hose to each injector instead.

Putting a throttle body on top of your port injection setup to meter the airflow is a bit more straightforward. There are several throttle bodies available which bolt directly in place of a Holley four-barrel carburetor. Another popular option is the intake elbow. This is a sort of adapter with a 90 degree bend in it, with the bend being added because many throttle linkages have difficulty clearing the large, flat mounting surface on a carbureted manifold. Off the shelf intake elbows usually accept a throttle body from a late model injected V-8 like a Ford modular motor, and attach to Holley four-barrel flanges. It's relatively straightforward for a fabricator to build one for a more unusual throttle body or carburetor using metal plates and a section of industrial pipe, or make one from sheet metal.

similar to an old-fashioned single-plane intake. These intakes give a good steady flow into the plenum that the ECU can easily measure, and they are cheap to mass produce. They are also a good deal easier to build for turbocharged or supercharged engines as there's only one throttle body to seal from leaking.

Although a straight runner is ideal, a manifold designed for port fuel injection will tolerate a lot more unusual shapes in the runners than one meant for a carburetor. It will also let you get away with designs that would create fuel distribution nightmares in a carbureted motor, such as runners that flow straight up, a ring-shaped plenum circling the cylinder heads, or other shapes that would make fuel drop out of the mixture. Some designers have taken advantage of this design freedom to create variable runner length manifolds, usually by adding extra valves in the manifold to let air through different passages, sort of like a trumpet. On rare occasions, you'll see manifolds that use sliding sections of runners like a trombone.

Independent Throttle Body Manifolds—An independent throttle body manifold has a separate runner for each cylinder, each with its own throttle body. There may be a plenum that supplies air to all the throttle bodies (this would be necessary on a car with forced induction) or the throttle bodies may simply each have an air filter or air horn on the end. Independent throttle body manifolds let you fine-tune the effective runner length by substituting air horns of different lengths. They offer the potential for immediate throttle response. However, this type of intake tends to be rather expensive, and they can be a bit more challenging to calibrate an ECU for without some experience.

Throttle Body Size—The throttle body (or throttle bodies) needs to be large enough to deliver enough airflow to the engine. However, making the throttle body too large can make the car tricky to drive at low speeds since a small amount of opening will let in a huge amount of air, more of less causing you to go from closed throttle to effectively full throttle, with little use of the actual throttle. If your engine happens to be popular for racing and runs fuel injection, you may be able to compare notes with fellow enthusiasts and find out what size throttle body will give you a good mixture of drivability and power. If you go off the beaten path, a bit of math will get you a good idea of the size you need (see sidebar above).

One way to determine if the throttle body you already have is costing you power is to measure the pressure upstream and downstream of the throttle body. If you're running an aftermarket speed

---

Determining Throttle Body Size

First, you'll want to calculate the airflow needs, in cubic feet per minute (cfm). You'll do this without factoring in turbo boost. If you have the engine size in liters, multiply it by 61 to get cubic inches (cid). Then use this equation to get the maximum airflow. If you don't have a value for volumetric efficiency, figure on around 80–90%.

$$cfm = (maximum\ rpm \times VE \times cid) \div 3456$$

If you have a throttle body rated in cfm, you might think this formula would give you the size you need for your throttle body, except for one thing: These throttle bodies are usually rated at a pressure drop of 1.5 inches of mercury (in.Hg), and, well, you don't usually want a pressure drop of 1.5 inches of mercury across your throttle body. That's like losing nearly 3/4 of a psi of boost. Multiplying the calculated cfm by 1.7 will give you a more acceptable size for a performance engine.

If your throttle body is not rated in cfm, you can use a rule of thumb to aim for a minimum airspeed of 300 feet per second (fps) through the throttle body to keep the engine drivable. So you can get the throttle body area you need by taking your cfm figure, and getting the area in square inches needed with this equation:

$$maximum\ area = (cfm \times 2.4) \div 300\ fps$$

Remember, this is just a rule of thumb, and some engines may be able to get away with larger throttle bodies. A progressively opening throttle body that opens faster the further you push down the pedal can help tame oversized throttle bodies. Also keep in mind that a drag racing engine that's only run at idle and maximum power will have fewer problems with a too-large throttle body. Road-race and autocross engines are more likely to have problems with too much throttle body.

---

density ECU, the MAP sensor will already give you the downstream measurement. The bigger the pressure drop, the more the throttle body is holding the engine back. The pressure drop can be like losing boost on a turbo car. If you're losing 0.75 psi across your throttle, you definitely have too much of a pressure drop. On the other hand, if your pressure drop is only a tenth of a psi, a bigger throttle body would probably do more harm than good, or at least be a gratuitous waste of money.

Idle Air Control Valve—In addition to the throttle body, there's often another valve that lets air into the engine: the idle air control valve (IAC). This is like a smaller throttle controlled by the ECU to control the idle speed. Some of the earliest designs were plain on/off valves that opened when the engine was cold and closed when it was hot. A

Most factory IAC valves attach to the throttle body, such as this one on a 5.0-liter Mustang.

If you're installing an IAC valve on an engine that didn't have one, the aftermarket offers a variety of fittings to make this easier. This IAC body from DarkStar Media accepts a Jeep IAC valve. One of the hose fittings goes upstream of the throttle body, the other one to the intake plenum.

Cross section of a stepper-type IAC valve and its housing. The plunger extends and retracts to allow air into the motor.

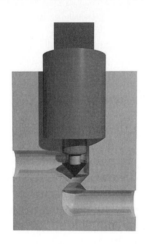

few other early designs used a thermal element that would open when cold, but an electric heating element caused parts in the valve to expand and close as the engine warmed up. More recent designs include the *pulse width* modulated valve or a *stepper motor valve*. The stepper motor is pretty much what it sounds like; a computer-controlled electric motor that moves in little steps. The pulse width modulated valve usually has a single electromagnet. The ECU turns the flow of current through the electromagnet on and off to regulate the current amount, with longer amounts of "on" time holding the valve open further (in most cases—a few valves have two electromagnets, one for opening and one for closing, and an even smaller number use current to close the valve).

Some recent intake designs take the idea of the computer controlling the idle speed a step further and use the ECU to control the throttle body itself. This is known as a *drive-by-wire* throttle. This lets the ECU both control idle and close the throttle to provide traction control. Some recent cars such as the GM cars with the newest version of OnStar even have a remote shutdown feature where an OnStar operator can reduce a car's power remotely if the car is reported stolen. It's a rare aftermarket ECU that can control a drive-by-wire setup, particularly since during a mistake in programming one could pin the throttle open. If you have a car with a drive-by-wire throttle and your ECU of choice cannot control this, converting to a cable-operated throttle body is often your best bet. Some higher-end standalone ECUs do have what it takes to control these, however. If you need this feature, do your research and make sure it will control your throttle.

That sums up the core of the mechanical side of things. From the fuel system to the throttle body, into the intake manifold, then BAM! Now on to some basics on the electrical side of things.

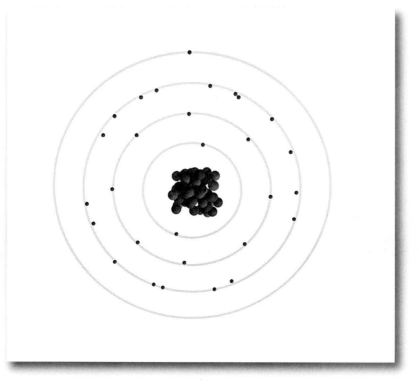

A copper atom, with electrons orbiting a nucleus. The outermost electron is only loosely attached and free to drift off the copper atom onto other nearby copper atoms, letting electricity flow through copper.

Chapter 3

EFI Wiring

Now that we've covered the mechanical components of EFI, the next step is wiring them up to the ECU. While you won't need to know enough about these to design an ECU from scratch, it helps to know a bit more than the smoke theory of electronics. (This tongue-in-cheek theory states that electronics are powered by smoke. The proof is that the device usually no longer works if you see smoke escaping from your wiring or the device itself, the smoke has escaped, the electronic lifeblood draining away.) Knowing what electricity does and how to build a sound wiring harness can help avoid letting the smoke out. An understanding of electricity will also help you get an idea of how the sensors we'll cover in the next chapter are going to work.

Electricity and Electrons

The basic idea behind electrical current is that not all of the electrons that circle the atoms in a metal are tightly attached to the atoms. Instead, some of the electrons can move from atom to atom, and it's possible to make these electrons circulate through a wire in one direction instead of drifting about randomly. The flow of electrons is called *current*. The electrons move from the negative side of a circuit to the positive side, so the direction of the current is actually the opposite of the direction where the electrons are moving. This may be confusing, but it's because scientists had named positive and negative before they figured out the part about electrons. The electrons must have a complete path of material that can conduct electricity for current to

flow; physically disconnecting the positive and negative side of the circuit stops the current.

Voltage is sometimes called electromotive force, because it's the force that causes the electricity to move. More voltage will push more current through a circuit. On most cars, power circuits run at 12 volts (and in practice can get as high as 15 volts, but it's still called a 12-volt system) and sensor circuits run at 5 volts. Occasionally you'll see exceptions; some older cars use 6 volt electrical systems, and it's likely that future cars will use higher voltage for some systems. Cars with oddball voltage wiring systems usually need to be converted to a 12-volt wiring system for using current aftermarket fuel injection systems.

Resistance is another electrical concept that you'll see mentioned often. Resistance is the ability of a circuit to restrict the flow of current. If you have a device hooked up to a constant voltage source like a well-regulated automotive charging system, cutting its resistance in half doubles the current flowing through it. Most of the time your dealing with resistance will involve checking how to set up your injectors and calibrating sensors. However, another application of resistance you will have to watch out for is the short circuit, where the resistance of a part drops to nearly zero. Some examples include a damaged injector or a broken starter cable touching the chassis (which is connected directly to the other end of the battery). Short circuits can let so much current through a circuit that things overheat, melt, and yes, let the magic smoke out.

Too much current can do serious damage to electronics. This Honda ECU had a capacitor fail and short out, drawing in enough current to burn the circuit board.

Capacitors store electric charge and can filter out noise.

Turning off the current running through a coil of wire creates a voltage spike called flyback.

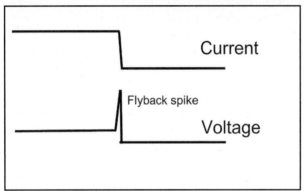

Current

Flyback spike

Voltage

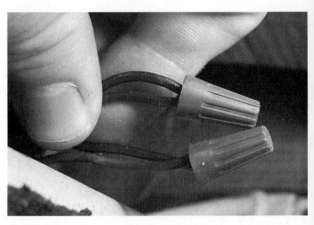

Wire nuts are likely to fall off from vibration and have no place in automotive wiring.

Both too much voltage and too much current can kill electronics, so you'll have to watch out for both. A mistake made during installation can cause either one; you can accidentally hook up a 5-volt device or ECU input to a 12-volt wire to give too much voltage, or wire in a device that draws too much current for the switch or computer controlling it. We'll get into how to avoid these potentially costly mistakes later.

Tools and Wiring

I once owned a Triumph Spitfire that had degenerated into a wiring nightmare by the time I bought it. While thirty-year-old wiring in English sports cars has a pretty rough reputation, the factory issues looked pretty trivial next to what damage previous owners had inflicted on the wiring. This car had wires routed where they rubbed against the hood hinges, several inches of the insulation was burned off in multiple spots. One pair of wires was only held together by bare ends of the wire twisted together without so much as electrical tape to protect it, and several stretches of the wiring had been replaced with lamp cord. And there were only three fuses in this entire copper rat's nest. The high beam switch caught on fire the first time I switched on the headlights. Careful attention to wiring could have prevented

this sort of trouble. Okay, ANY attention to the wiring might have prevented this mess!

Like most jobs, proper automotive wiring requires using the right tool for the job and the right materials. Even if you've learned wiring for other sorts of jobs, automotive wiring has its own requirements. For example, some techniques that would be perfectly up to code for home wiring are downright dangerous when exposed to the vibration, heat, and other things you have to deal with in a car. The wire nuts used to twist together two strands of wiring around the house are easily the worst offender. Too much electrical tape is also not so great. Use it in moderation, and make sure it's rated for the temperatures it will see. Don't buy the cheap stuff.

Basic Hand Tools

You'll need a couple different tools for wiring up your car unless you're able to buy a complete, ready-to-run wiring harness for your EMS and car or a

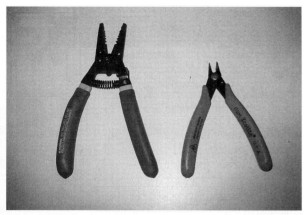

You'll need a good wire cutter and a wire stripper for any rewiring job.

A specialized set of crimpers for Weatherpack connectors makes assembling the connectors easy. You can crimp Weatherpack connectors with pliers or an ordinary crimper, but that would make it much harder to get a reliable connection.

plug-and-play system that reuses your factory harness. And even then, it's good to have a set of wiring tools handy in case you need to change something about an off-the-shelf harness or the stock wiring. Most of the wiring harnesses you'll see with an aftermarket EMS are what are called "flying lead" harnesses, with an ECU connection on one end and the other end of the harness unconnected so you can splice your own connectors on.

Two basic tools are a wire cutter and a stripper to trim wires to length and remove the insulation. Even a cheap wire cutter or stripper will do the job, although better quality tools will cost a bit more. Some of the cheapest wire stripper tools are less than satisfactory; however, a good one will be able to grab the wire from both sides at the same point, while cheap sheet metal cutters often have the sides go out of alignment and make the job more difficult.

Inductance, Flyback, Capacitors, and Other Parts

There are more electronic concepts that are helpful in understanding how certain circuits work. These can be a bit more confusing than voltage, resistance, and current, and they're usually more important to people designing electrical circuits than those installing EFI. But it can still be important to know how they work even if you aren't working for an EFI designer or trying to add a new custom circuit to a homemade MegaSquirt. And when you're putting together your wiring harness, you are, in a somewhat simpler sense, designing a circuit.

Inductance—First one up is inductance. You're probably familiar with how current running through a coil of wire creates a magnetic field in an electromagnet. This magnetic field doesn't just pick up iron filings; it can store energy as well. When the current is turned off, the energy in the magnetic field creates a voltage spike, and this spike is larger than the voltage that originally created the magnetic field. This effect, called *flyback*, creates a lot of electrical noise. Ignition coils and fuel injectors both create large amounts of flyback; other devices with coils in them, such as relays, can also do this. Even a normal wire can create a small amount of flyback, since electricity traveling down a wire creates a small magnetic field around it.

Capacitors—Another component you'll often see is a capacitor. The main function of a capacitor is that it stores an electrical charge. Capacitors are kind of like electrical sponges, capable of quickly charging, storing, and also quickly releasing energy on demand. Capacitors can also be used to cut down on electrical noise, because when voltage suddenly rises, the capacitor grabs some of the extra charge now flowing through the wire. If the voltage suddenly drops, the capacitor releases some of its stored energy, in a sense smoothing out the voltage on the circuit. Capacitors are also called condensers. If you've worked on old-fashioned points distributors, the condenser inside them is a type of capacitor.

There are a couple of other devices you'll sometimes see mentioned that we'll go over real quick here. Diodes are electrical one-way valves. The typical diode has a stripe on one end. Current can flow through the unstriped end to the stripe, but not the other way around except for a particular type of diode called a zener diode that lets current flow bi-directionally once it exceeds a certain voltage. Transistors are another component you'll often hear about; these act like electrical switches that get turned on or off by an input voltage, sort of like a low-current, fast-acting relay, but not quite.

Most of the time when installing an ECU, you won't have to worry about adding these parts. But it can help to know what they do. Okay, go ahead and forget most of that now, just know where to find it if you need it.

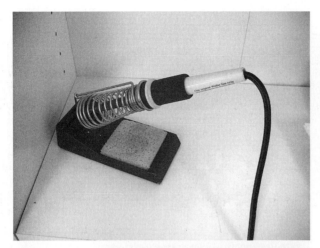

A Weller 60-watt soldering iron, suitable for working on both harnesses and circuit boards.

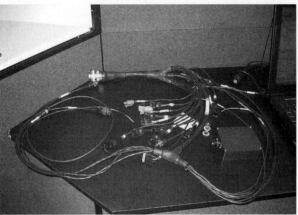

To make a reliable harness, use the right materials. This harness from EFI Technology takes reliable construction a step further than production car wiring, with mil-spec connectors and other careful attention to detail.

Solid-core (top) vs. stranded-core (bottom) wire. Solid-core wire can easily break from vibrations. Use stranded wiring to make a reliable wiring harness.

If you're working with crimp connectors, you'll need an appropriate crimping tool as well. Cheap sheet metal crimpers seldom work very well. A sturdier tool that resembles a set of pliers typically can make effective, reliable crimps without too much effort. Ratcheting crimp tools require specific dies for each type of connector. But if you're crimping Weatherpack connectors or other specialized items, a tool specific to the style of crimp connector you're using is sometimes the best option.

You can use a soldering iron for working on circuit boards and for soldering wires together. The wattage rating of your soldering iron matters. A cheap 15-watt iron will take a long time to heat up the work you are attempting to solder. A 60-watt iron is good for general purpose wiring and for typical circuit board work. Bigger tips for wiring work, smaller for circuit board work.

Types of Wire

While most wires look alike except for their thickness and color, there are several important differences that aren't as obvious. The wrong wire won't stand up very long to the heat and vibration of an engine bay. You will need wiring with the right kind of core, the right kind of insulation, and the right thickness.

While most wire has a copper core, there are two different ways to build the wire core. Solid wire is exactly that—a single piece of copper running through the wire. Stranded wire has a core that looks like a rope made from many thin copper fibers. Solid-core wire is meant for household wiring, and it doesn't belong in an automotive wiring harness. Vibrations put more stress on a solid core than a flexible bundle of strands. Over time, solid-core wire will fatigue and crack if you use it in an automotive wiring harness.

Insulation Types—Automotive grade wire comes with four types of insulation. GPT insulation is for chassis wiring and is rated at -40 degrees to 80 degrees C (all the ratings in the official specs are in metric). This doesn't really belong in an engine compartment; it's for tasks like wiring up your taillights, away from the engine bay. The underhood rated wiring grades are TXL, GXL, and SXL (sometimes called SGX). These are all rated for -40 to 125 degrees C. TXL has the thinnest insulation, but it's great for general purpose wiring and keeps your wiring bundle from being the size of your leg when it's complete. SXL is the really thick insulation for battery cables, and GXL is somewhere in the middle. Stick to the three XL types of wires for your engine harness, unless you happen to get a deal on some military or marine grade wire that has the same sort of temperature resistance.

Automotive grade wiring comes in many colors. If you're not buying a ready-made harness, you'll want to take advantage of this and not build a harness entirely out of one color of wiring. Use different colors for different circuits, and take notes when you wire the car up. Although black is the goes-with-anything color, you'll have a much easier time troubleshooting your installation later if you have used different colored wires for each circuit.

Wire Size—Non-metric wire is sized by gauge number. The larger the wire, the smaller the number. The number itself doesn't really match any specific measurement of the wire. If the wire is too thin for the amount of current it carries, it can overheat and burn out. The chart gives recommended wire sizes for a typical run in a car (15 feet or less). The examples are rather general; you may run across devices that draw much more or less current, so check the specifications of the actual device you're installing for how much current it draws. If you don't have any specifications, you can usually get the current draw by dividing the car's voltage (go with 15 volts for safety) by the device's resistance in ohms. A 16-ohm injector, for example, draws just under an amp of current. A device may draw a bit more when it starts up.

Very small wiring, like 22 gauge, can pose its own problem. Even when it is able to carry enough current, it has very little mechanical strength. With very low current applications, you sometimes may want to go with a larger wire for durability.

Planning a Harness Layout

Some places in an engine compartment are not the best places for your harness to go. While a wire touching an exhaust manifold will obviously not last, there are several other ways you can run wires that are nearly as bad but with far more subtle dangers. Physical damage is not the only threat. Wires can act like radio antennas and "tune in" electrical noise that can either give the ECU inaccurate readings or, in extreme cases, scramble its programming.

Physical Hazards—Most physical dangers are pretty straightforward to anticipate. The wires need to be kept away from moving or hot objects and tied back with an appropriate support to make sure that vibration or air current can't bring them into contact with these hazards. Don't forget such things as throttle linkages, transmission kickdown linkages, or retractable headlight doors when making a list of moving hazards; it's not just the accessory drive belts, fans, and pulleys that you have to deal with. Also, when you're running wires from the engine block to the firewall or fenders, assume the engine may move a couple inches when torque is applied—leave enough slack in the harness.

Wire Gauge Sizing Guide		
Wire Gauge	**Amps**	**Example Device**
8	32–40	Usually used to power several devices on one circuit
10	28–35	Very high-powered ignition
12	18–30	Racing fuel pump
14	12–20	Injector bank, headlight, typical cooling fan
16	8–13	Typical fuel pump or high-performance ignition
18	6–10	Nitrous solenoid
20	4–6	Exterior signal lights, single injector
22	2–3	Sensors, small interior lights

Letting part of your wiring harness come in contact with the spark plug wires is asking for trouble.

Electrical Noise—Electrical noise is hard to pinpoint, but you can keep it to a minimum by identifying the potential noise sources and keeping critical wiring away from them. One repeat offense I've seen in badly designed wiring harnesses is letting a spark plug wire touch, or even get too close, to a wire leading to the ECU, particularly a 5v signal wire like a sensor signal. While spark plug wires have insulation to cut down on noise, don't trust that to stop it from generating currents in a wire it's physically touching or near. Under no circumstances should you ever allow any wire to the ECU to come in contact with a high-voltage spark plug wire unless you are deliberately using it as a sensor to pick up the current in the spark plug wire. However, it's a rare application that uses a wire like that.

The rest of the ignition is also rather noisy. Some parts need to have wires connected to them—coils need wires to control them, and there may be a sensor in the distributor, for example. In this case, you should only bring the wires needed for the device's operation near it. Routing a sensor wire behind an ignition coil is likely to draw in a lot of

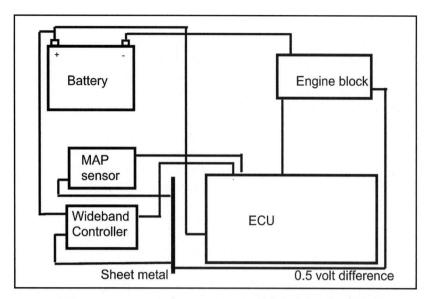

Ground offsets can happen if you ground different parts of your system to different parts of the car.

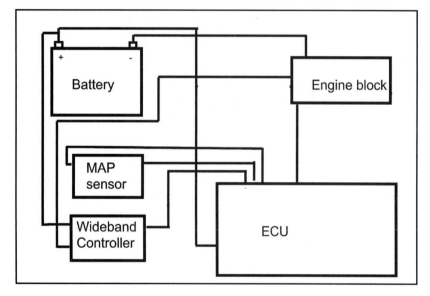

A few changes to the wiring diagram in the previous illustration can vastly improve sensor readings.

noise. Keep low-voltage sensor signal wiring far away from high-voltage noise sources and you'll have much less chance of having weird issues to chase down later.

If you're wondering just how much noise your engine electronics put out, you can hear the noise with a cheap AM radio. Tune it to a setting where it can't pick up any local stations, and sweep it around your engine compartment. The radio will tune in the electrical noise and you can hear it get louder when the radio approaches a noise source.

One other way you can minimize noise is by keeping the high-current wires (wires that drive the

injectors, fuel pump, and the like) separate from the signal wires. This cuts down on "cross talk" effects where changes in one wire's current creates changes in another wire. As a noise source, this generally has a lesser effect than high-voltage sources as mentioned earlier, so put the priority there.

Grounding—Critical. Enough said? Not hardly. The ECU and its inputs (sensors) must be properly grounded. If you're filling a show car with neon lights, you could ground the lights to any hole you drilled in the chassis and they'd still work. However, that sort of grounding is about as appropriate for a standalone ECU as those 50 lb of neon would be in a Formula Vee race car. Computers are precision instruments, and precision instruments require precision grounds. It's best to ground the ECU to the engine block or the negative battery terminal. Engine block grounds are less affected by corrosion than the battery terminals, while the battery terminal gives it a straight shot at the battery. While some in the EFI community argue over the merits of these two different locations, the two camps agree that you should never ground an ECU to any other place.

This is the Garbage In, Garbage Out principle. The trouble with just grabbing any chassis ground is that, while in theory all grounds are at zero volts, in practice this isn't the case. There is a little bit of resistance in ground paths, and if you run enough current through a ground, the voltage at the ground point will go up. If the ECU thinks 0v is different from what the oxygen sensor thinks 0v is, they have a different reference voltage and that throws their entire scale off and their entire ability to communicate and share information off. Let's say you ground the ECU at the battery, and ground your wideband O_2 sensor on the frame near the rear of the car. There's going to be much more resistance in the wideband's ground than there is in the ECU's ground. For the sake of discussion let's say that introduces a 0.25v offset due to this resistance. What the ECU thinks is ground (0v) to the wideband is actually ECU ground minus 0.25v (−0.25v), but the wideband doesn't know this. You wired this up as it's ground, and the wideband just thinks this is 0v. All of your readings from that wideband's output will now have a 0.25v error introduced in the reading. What's reading that output? The ECU is, and you might have a gauge displaying the air/fuel ration (AFR) to you in real time. And it will never have the right information to display to you unless you wire either one correctly right up front.

The type of wires used for grounding can be almost as important as where you ground the ECU.

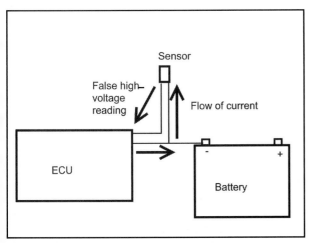

This ground loop could keep you from getting an accurate sensor reading.

It's better to have several skinny ground wires than one big fat ground wire. The reason is pretty complicated, but the short version is that a collection of smaller wires will have less inductance than one big wire. The resistance of a wire with high inductance actually goes up if the current changes suddenly—like when you turn an injector on or off. So while a single fat wire may have the same resistance as six thinner wires when measured with an ohmmeter (if you can find an ohmmeter that will go that low), it will have more resistance when hooked to an actual, live ECU. Also, some ECUs keep their signal grounds and their power grounds separate, and combining them before they reach the chassis ground can hurt the ECU's noise resistance.

Another thing to avoid in grounding an ECU is ground loops. Grounds should converge as they flow back to the battery, with the sensors grounded to the ECU, the ECU and other components grounded to the engine block, and the engine block grounded to the battery. If a sensor ground goes to both the ECU and the engine block, for example, current could flow from the ECU to the sensor ground and then to the engine block. This flow of current would interfere with the sensor reading.

ECUs have metal cases, and on some ECUs, the case is grounded. If you have an ECU like this, be sure to mount it so that the case is not in contact with any metal on the car, even if you have to use plastic screws to accomplish this. Grounding the case on the ECU to the chassis will create ground loops, and there have been a few cases of ECUs like this acting up where mounting them on rubber pads fixed the problem.

Dealing with Noise

Sometimes, even when you've carefully built a wiring harness to avoid sources of noise, you still have signs that it's there. This is particularly common if you have installed EFI in an older car where the designers never expected to see any electronics more sophisticated than an AM radio. Common signs of this include battery voltage readouts that go haywire, sensors readings that give you static, and sometimes ECUs just plain misbehave.

The first thing to do is double-check for noise sources that don't need to be there. Make sure you're using resistor type spark plugs and good quality plug wires, as this can often make quite a difference. And no, this doesn't cost you by making the spark significantly less powerful. Also, double-check for wires running near sources of noise mentioned above.

When the noise comes in on a single sensor, check the sensor and everything between it and the ECU. The connector to the sensor can be corroded, or you may need to replace the wire to the sensor with a shielded wire to keep it from picking up noise. The sensor may have gone bad, you might have a noisy alternator, generator, fan, coil, or ignition box introducing noise somewhere along the line. Find it and correct it.

Joining Wires

There are several ways to splice wires together when you're making a harness. Two good ways are crimp connections and soldering. In theory, crimping has an advantage over soldering in that solder soaks into the strands of wire and makes it act like a solid core wire. However, I've seen solder joints in automotive wiring harnesses that are over 40 years old still holding together quite well. Either method can produce a great wiring harness if done right, and either one can break if done wrong. There are also a few ways to join wires together that will NOT produce a safe connection at all.

Vampire Clips—Vampire clips are a type of quick-connect clip that sometimes comes with aftermarket electronics. These are designed to pierce an insulated wire and tap into it with a metal part that looks like vampire's teeth. They install very quickly and often take just a pair of pliers to put them on. Unfortunately, this can often cut through several strands of the wire, weakening it. They also do not protect the wires from corrosion. They're not the best fasteners to use for serious, durable electrical work in your car. Vampire clips can bite in more ways than one.

A pair of taps, also known as vampire clips. These can weaken the wires they clip to and are not as reliable as soldered or crimped connections.

Crimping two wires together with a butt splice connector.

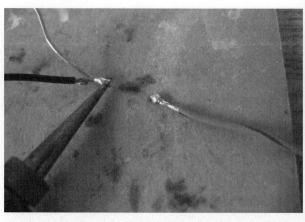

Soldering two wires together. It helps to first apply solder to each individual wire. Tin the wires enough and you can join the two together by touching them with a soldering iron where they meet.

Wire Nuts—Perhaps the worst repeat offender in bad wiring jobs is the wire nut. It is often used in home wiring, where it lets you make a very quick connection with minimal tools. And it's perfectly safe in a light switch box in the wall. However, a light switch box isn't supposed to be bouncing over potholes at 60 mph, and the wire nuts are meant to be used with solid-core wiring. When you put them on stranded wire and send them through the Corkscrew at Laguna Seca, wire nuts can, and do, vibrate off the wiring.

Stripping the Wires—So, back to soldering or crimping. Joining wires, regardless of the method used, starts with stripping the insulation off the wires. While you can do this with a knife or a wire cutter, it's a lot easier with a purpose-built wire stripper. This tool will have a set of holes for different wire sizes and makes quick work of the insulation. Good quality tools are just as important with wiring as they are with wrenching. If the connection is going under the hood, you'll also want to put a length of heat shrink tubing over one of the wires to be joined, long enough to cover the splice. If you're soldering the wires, make sure

they're well away from the area to be soldered so the heat from the soldering work doesn't shrink the wrap before you are ready.

Crimping—If you're crimping the connectors, your next step is to simply insert the stripped end of the wire into the barrel of the crimp connector and crimp it with a crimp tool. There are several types of crimp tools on the market. The ones I've found work best resemble a pair of large pliers with an indentation on one side and a prong on the other. These work great for general purpose crimp connectors. Cheap sheet metal crimpers often don't have enough strength to make a good crimp, while specialized crimpers with elaborate crimping dies are usually specialty tools that work with only a particular wiring size and particular fastener.

If you're soldering the wires together, you'll need to make sure the wires get hot enough. You need to heat the wires to the point the solder is actually drawn to the wire. If the solder seems to ball up and wants to stick to itself more than to the wires, you'll end up with a cold joint. Cold joints tend to break easily. It helps to apply a little solder to the iron before touching it to the wires as it will transfer heat to the wire better. I like to strip both wires, point them at each other overlapping one another, and twist them together so that the wire continues to flow as one long wire once twisted together. Then heat the exposed copper and feed solder as it wicks into the joint.

Protecting the Connection—Once you have the joint in place, unless you have put an insulated crimp connector over the wire and are running it inside the passenger compartment, you'll need to cover up the joint. Even insulated connectors need water sealing in an underhood environment. And it

Heat-shrink tubing protects your newly joined wires from air, water, and vibration. It's a good idea even if the wires are not under the hood. The connectors in this picture come with heat-shrink tubing already in place.

can still be a good idea to seal connectors under the dash to protect them from corrosion. Heat-shrink tubing works very well for this purpose. If you forgot to put the tubing over the wires before joining them, you can use brush on liquid electrical tape, although this is a bit messier. Heat-shrink tubing or liquid electrical tape also protects connectors from vibration, so it can pay off to use this even in a dry environment. You can also buy crimp-type connectors that already have a small length of heat-shrink tubing on them.

Connectors—In addition to splicing the wire, you'll often find you need to put connectors in it, either to make sections of the harness removable or to attach sensors, injectors, and the like. Most of the connectors you'll see in the aftermarket are GM-style Weatherpack or Metripack connectors that are fairly water resistant. They are designed to have their pins crimped in place with a special crimping tool. In a pinch, you can solder them on or use pliers, although this isn't ideal. You can also buy pigtails that have short lengths of wire already crimped on, ready for you to splice into your harness. These are often a good choice if you don't have the specific crimping tool the connector needs.

Fuses and Relays

Fuses and relays both protect wiring and circuits in different ways. Fuses contain short lengths of metal that melt if you run too much current through them. Any time you are adding a new circuit to your wiring, you should give it a fuse as close to either the positive battery terminal or whatever power source

you are tapping. The fuse should be sized just slightly larger than the most current the circuit will carry. Note that electric motors draw more current when they start than when they run. Usually the manufacturer of a fuel pump, fan, or similar device will be able to suggest what size fuse should be used in its power wire.

Fuses—At a minimum, you'll want fuses for your fuel pump, one or two fuses for your injectors, and a fuse for your ECU. It's a good idea to have a larger fuse for all your EFI components with these other fuses getting their power from the main fuse. If your car is already wired for EFI from the factory, it should have these. If not, there are many sources for fuse blocks to add fuses. Auto parts stores often have cheap and effective blocks, while boat supply stores often carry more rugged options.

Relays—Relays are mechanical switches controlled by electromagnets. These let you turn on a device that draws a lot of power using a flimsy switch or electrical circuit that can't handle the full power of what it is turning on. For example, if you have a monster 40-amp cooling fan, you don't want to run all 40 amps through your ECU. Instead, you can have your ECU turn on a relay that requires a fraction of an amp to activate, and the relay handles the 40 amps.

The typical relay has four terminals, although you'll also see relays with five. Bosch came up with a numbering system for these terminals that's become an industry standard. Pins 85 and 86 power the relay coil; one of the pins needs to be connected to a 12-volt source and the other to ground in order to activate the relay. You can turn the relay off by either disconnecting it from 12 volts or disconnecting the other end of the coil from the ground. The electrical power is supplied to pin 30. When the relay is active, it sends power from pin 30 to pin 87. Relays with a fifth pin call the fifth pin 87A. They are set up so that when the relay is not activated, current flows from pin 30 to 87A, and this turns off when the relay is activated.

A few relays will have other numbering systems, but if they do, it's usually marked on the case. Relays typically use this symbol, and you can use the symbol to identify which connection does what by comparing the terminals on the relay's schematic to the Bosch numbers.

Diagnostic Tools

Multimeters—In addition to tools for connecting the wires, you might want a few tools for examining the electronics. The most important tool to have is a digital multimeter. This tool uses two probes to measure voltage, resistance, and current. Some

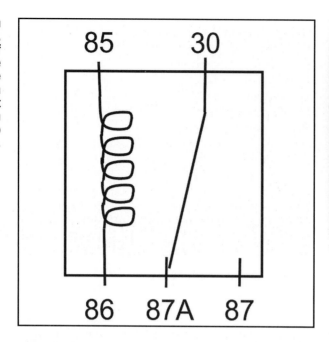

Standard Bosch relay numbering. The coil half controls the switch. The switch may open and close or it may switch between two settings.

If you're doing any ignition work, you need a timing light. This Equus light has a digital dial-back adjustment and works with almost any ignition system out there.

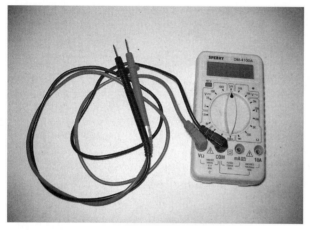

A digital multimeter is a must-have tool for trouble-shooting your electronics. You can get by with a low-priced one, but spending more will get you a tougher model.

multimeters can also measure frequency, inductance, or capacitance as well. The voltage-testing mode is the one you'll probably use most often. It can be used to find good power sources, test some types of crankshaft position sensors and distributors, make sure relays are working, and a lot more. The resistance tester is also useful—you can check what type of injectors you have and check the operation of some types of sensors with it. Occasionally you'll find uses for their other functions, too.

One drawback of a multimeter is that it often doesn't react very well to rapidly changing signals. For example, if you have a 12-volt source that sometimes has short spikes up to 24 volts, you probably won't be able to measure the pulses. The meter will just sit at 12 volts, maybe sometimes flickering. Sometimes the frequency meter settings can determine that a voltage that should be changing is changing, such as injector inputs.

The $30 digital multimeters work reasonably well, but they're usually a bit on the fragile side and may not be quite as accurate. Spending more on a professional unit from Fluke or another high-end supplier will get you a unit that's designed to hold up to the sort of rough handling that electrical contractors and mechanics frequently encounter. High-end multimeters often add other capacities, such as measuring peak voltage. Most of us don't need to have a high-end unit, unless you're just that kind of person.

Old-fashioned analog meters are often less accurate than digital meters. They are useful for checking a rapidly changing signal, however. If a device's output flickers rapidly between 6 and 12 volts, an analog meter's needle will often let you know you've got a moving voltage and you may be able to find the average voltage by eyeballing it. But if you can get only one multimeter, make it a digital model.

Test Lights—There are also a few inexpensive types of test lights that you may have a use for. A basic test light will simply show if voltage is present at a certain point in the wiring and is handy for quick checks. A noid light is a type of test light that can react to quick changes in voltage and is useful for such tasks as determining if injectors are getting pulses to fire.

Timing Lights—Timing lights are also a type of test light, although they're a bit more complicated and expensive. These systems use a pickup coil wrapped around a spark plug (or rarely with older timing lights, a spring of wire fitted between the plug and the spark plug wire) to sense when the spark plug is firing. You point the business end of the timing light at the crank pulley. The timing light fires very quick bursts of light that make the

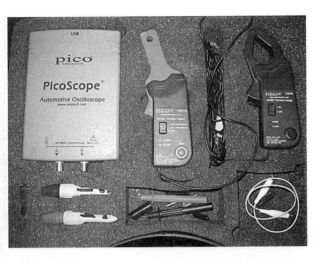

A USB oscilloscope. This one plugs into your computer and lets you record signals for examining later.

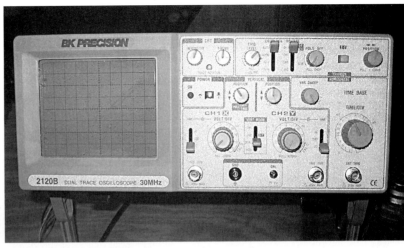

A larger, lab-style oscilloscope.

pulley appear to be frozen in time. Using the timing light and marks on the pulley, you can determine where your ignition timing is.

One common feature on timing lights is a dial-back knob. This lets you shift the time when the light fires. You use the knob in combination with a zero degree marking on the crankshaft pulley, which lines up with a point on the block when the engine is at top dead center. Adjusting the knob lets you measure the amount of timing advance. For example, if you have to adjust the knob to the 20-degree marker to make the marker on the pulley line up with your TDC mark, your engine is running 20 degrees of timing at that point.

Before buying a dial-back timing light, make sure it's compatible with your ignition type. Some of the lights can only work with a distributor. And if the coils are mounted on the spark plugs without any spark plug wires at all, it makes it difficult to use a timing light, but not impossible. Often you can detach the number-one coil and connect it to the plug with a spark plug wire, giving you a place to clamp the timing light pickup.

The 'Scope—For the hardcore electrical geek in some of us, the oscilloscope is the ultimate

diagnostic tool. Like a multimeter, an oscilloscope measures voltage or current. Unlike a multimeter, it tracks this over time and displays a graph. Oscilloscopes run quick enough that they can view the individual pulses firing ignition modules, or other signals that last for mere milliseconds (or nanoseconds, or picoseconds even). They can be used to measure some of the inputs and outputs to the stock ECU and/or ignition system components, so you can wire up your new standalone to duplicate its signals without some of the the usual trial and error.

There are several different types of oscilloscopes. The large cabinet mounted ones you see in labs can cost a small fortune, but you can pick up a cheap unit that plugs into a laptop's USB port for under $200. Even a cheap unit is often perfectly suitable for basic EFI troubleshooting. Note, before trying to trace any high voltage signals, read up on your scope's manual—you can damage your scope without the proper precautions!

Now that we've covered the basics of wiring, it's time to have a look at the more electronic side of engine management, the sensors.

Chapter 4
Sensors and How the ECU Uses Them

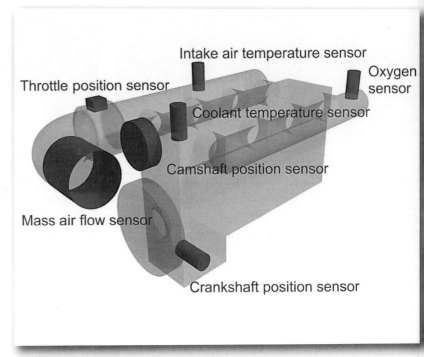

Intake air temperature sensor

Oxygen sensor

Throttle position sensor

Coolant temperature sensor

Camshaft position sensor

Mass air flow sensor

Crankshaft position sensor

This diagram shows many of the sensors you're likely to see in an engine, with the sensors in typical locations. Some installations do not have all the sensors from this picture, while others have a couple sensors not shown.

Okay, so you've got your fuel injectors wired up, ready for the ECU to operate. Now the ECU needs to know how much fuel to add. Your brain would have trouble figuring out what was going on around you if it wasn't connected to eyes, ears, or at least fingers to feel the outside world—likewise a computer is not going to have a clue what is going on without some input from a few sensors. So a fuel-injected engine needs a collection of sensors, and the ECU needs to know how to use these sensors. Not all ECUs use the same types of sensors, either, but there are several common sensors that you'll encounter on almost every aftermarket ECU available. We'll try and help you to understand what all systems will have in common, and to navigate the differences you might run into as well.

Some ECUs are very flexible and can be used with the sensors already provided from the factory on a fuel-injected car, while others require you to use a specific model of sensor generally offered with the ECU. You will need to verify during the planning stages of your EFI system which you have. Most systems that require a specific sensor use sensors out of the late 1980s and early 1990s General Motors parts bin, although there are a few exceptions. The sensors may be sold separately, included with the ECU, or even inside the ECU and connected to the engine with a hose.

Engine Position Sensors

The first and most critical thing an engine management system needs to know is how fast the engine is turning. If you are controlling spark or using sequential injection, the ECU will also need to know where the pistons are at any given time. The engine may use a single sensor for this task, or it may have two (or occasionally more) sensors working in tandem. These sensors usually use a trigger mechanism mounted on a wheel, which might have bumps or teeth on it, or slots or holes cut in it.

There are several different sorts of sensors that a designer could use for this purpose. The crudest would be to have a wheel with bumps on it, and a spring-loaded switch touching the wheel. This switch would turn on each time a bump touched it. Some of the oldest electronic ignition systems use just such a switch, known as *breaker points*, for their ignition control. A few early EFI systems use breaker points, too. Many aftermarket ECUs can work with breaker points. But since these switches can easily wear out and are not especially accurate, this isn't ideal.

Variable Reluctor Sensors—The variable reluctor sensor, sometimes called an *inductive pickup*, magnetic pickup, or magnetic reluctor, is a simple way to get a position sensor that won't wear out. These combine a coil of wire and a magnet, and are closely related to the pickups on electric

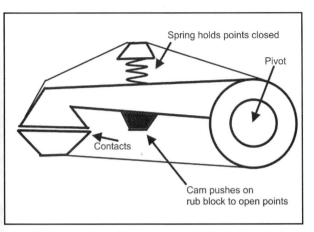

Breaker points are the oldest type of engine position sensor, found inside distributors up until the 1970s where other sensors became more popular.

The Pertronix Ignitor is an example of an ignition system that uses a Hall effect sensor. The ring at the right contains a set of magnets and attaches to a spinning shaft. A Hall effect sensor inside the module at left detects the magnets. Photo courtesy Pertronix.

The LS1 motor uses a Hall effect sensor for crankshaft position.

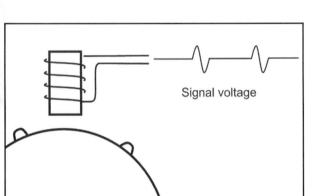

A variable reluctor uses a coil to generate electric current from changes in a magnetic field.

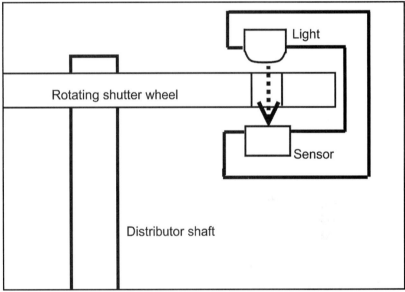

An optical sensor shines a light through a series of holes in a rotating disc.

guitars. In some arrangements, the magnets are on a spinning wheel, and create a current in the coil each time one of the magnets passes by the coil. On other sensor designs, the magnet is inside or next to the coil, and a tooth on a steel wheel next to the coil creates the electric pulses. The faster the wheel spins, the more powerful the pulses. A VR sensor can only detect the wheel if the wheel is moving; holding a magnet steady in front of it won't produce a signal.

Hall Effect Sensors—Hall effect sensors are magnetic too, but they're a more sophisticated type of magnetic sensor. They can detect a piece of metal in front of the sensor, whether the metal is moving or not. A typical Hall effect sensor contains a chip that acts sort of like a switch, turning on when there is a piece of metal in front of it, and turning off when the metal is gone (or sometimes the other way around). In some ways, a Hall effect sensor is more precise than a VR sensor. If you compare the outputs of the two sensors, you'll see that the VR

sensor is wavy (technically, it's a sine wave), while the Hall effect has nice sharp edges that makes figuring out the position a bit easier on the ECU. These sensors present a square wave signal. It's either on or off, with nothing in between.

Optical Sensors—Magnets aren't the only way to get a wear-free position sensor. Some engines use optical sensors that shine a light through holes in a spinning disc. A phototransistor on the other side of the disc detects when a hole passes in front of the light. From a wiring standpoint, optical sensors work in about the same way as Hall effect sensors, also presenting a square wave, on or off signal.

A designer can put a position sensor on almost anything that spins in time with the engine, but

The cam angle sensor on an early Miata sits at the back of the cylinder head and is used for both crankshaft and camshaft position information. It contains two optical sensors.

A cutaway MSD distributor shows its advance mechanism. The vacuum can at the lower left adjusts the position of the sensor, while the centrifugal mechanism in the middle of the distributor (arrow, underneath the sectioned rotor) adjusts the position of the reluctor wheel. Changing the springs changes its advance curve.

there are three positions that make the most sense. It can go in a distributor, on the camshaft, or on the crankshaft. The sensor can be known as a distributor pickup, cam angle sensor, or crankshaft position sensor depending on its location. Perversely, the sensor name isn't a 100% reliable predictor of where it is. While you won't see a cam angle sensor on the crankshaft, I have seen factory literature frequently call something a crankshaft position sensor when it was actually on the camshaft.

The earliest electronic fuel injection systems put the engine position sensor in the distributor. Distributor setups often (but not always) have one trigger per cylinder; this is especially typical on vehicles up through the mid-90s. In some cases they may have a second distributor pickup also, with more or fewer triggers.

Most carbureted engines, and a few older engines set up from the factory with fuel injection, use

what's known as an *advance mechanism* to control the timing. This changes either the position of the trigger wheel or the position of the sensor as the engine rpm and load change. Such a sensor can provide an ECU with the engine's speed, but isn't very useful at telling the position. If you are dealing with a distributor with an advance mechanism and would like to control ignition with a standalone EMS, you will need to remove and/or lock down (bolt or weld in place) the advance mechanisms to disable them, unless you want the ECU to just control the fuel. Other distributors intended for computer controlled timing, such as the Ford TFI or later versions of the GM HEI, don't have an advance mechanism and are pretty easy to set up for aftermarket EFI. Just make sure there is no (working) mechanical nor vacuum advance mechanism in place if you plan to use the distributor to control ignition on your engine.

Crankshaft Position Sensors—Many recent engines use crankshaft position sensors, often abbreviated as CKP sensors. These may be located either on the flywheel or on the front of the engine. A few engines even use the teeth on the flywheel ring gear as a trigger wheel, or put the trigger wheel somewhere on the middle of the crankshaft and have a sensor inside the block itself to read it. The sensor itself is often a VR sensor. Crank triggers are quite precise, because they are not affected by things like worn-out timing chains or belts which can introduce some "jitter" into the reading. If you're using batch-fire injection and a distributor or wasted spark ignition, you can get by without any more engine position sensors. But while a sensor on the crank can tell you how far a cylinder is from top dead center, it's not able to tell if it's on the compression stroke or the exhaust stroke. So if

Different engines use different trigger patterns. This is an example of the common Bosch missing-tooth crank trigger wheel.

Sometimes an aftermarket ECU may perform better with a different trigger pattern than the one on the stock engine. This Saturn features a homemade cam angle sensor setup using a missing-tooth wheel mounted on the cylinder head. The arrow is pointig to the sensor itself.

The throttle position sensor usually sits on the end of the throttle shaft, as shown on this 5.0 Mustang. Occasionally you'll see a TPS mounted remotely, connected to the throttle with a linkage. Using a linkage allows one sensor to fit a wide variety of throttles, but putting the sensor on the shaft is simpler and has fewer parts to break.

you're using sequential injection or coil-per-plug ignition, you'll need a bit more.

Cam Angle Sensors—That bit more is the cam angle sensor, sometimes called a CMP sensor or CAS. This spins at the same speed as the camshaft, and may be on the cam itself or in the distributor. If you are converting an older engine to use a crankshaft position sensor and cam angle sensor, modifying the distributor can be a simple way to add a camshaft position sensor. The modification you'd be after would be to remove all of the triggers from the distributor's trigger wheel except one to indicate top dead center (TDC) on the number 1 cylinder. This combined with a crank trigger wheel will provide all the information you need to trigger a sequential EFI and ignition EMS, provided your EMS supports the trigger wheel pattern and second trigger input from your modified distributor. As always, before you spend the time, check first.

Engineers have come up with a large, and sometimes inexplicable, variety of ways to arrange the teeth on their trigger wheels. And not all trigger wheels work with all aftermarket ECUs. The most common pattern is a distributor with one trigger per cylinder, or a crankshaft position sensor with half as many triggers as the number of cylinders. Sometimes this is paired with a single-toothed camshaft position sensor (the common dual-sync distributor is an example of this setup). This will work with most, but not all, aftermarket ECUs. Other common arrangements have a crankshaft position sensor with many equally spaced triggers, or a wheel with between 36 and 60 teeth, with one tooth (or two adjacent teeth) missing. This so-called missing-tooth wheel may also be paired with a single trigger camshaft sensor.

However, some manufacturers use very

complicated and creative designs for how they space the teeth on their position sensors. Subaru and Chrysler products since the mid '90s are two good examples. If you have an oddball trigger wheel setup and want to use an aftermarket ECU, check to be sure the ECU can use your trigger wheel(s) and sensors. If your ECU of choice does not work with your original wheels and sensors, however, this doesn't need to prevent you from using this ECU on your engine. Often you can install a crankshaft position trigger and sensor kit that attaches to the front pulley on the crankshaft and works with your ECU. And with some aftermarket ECUs, this is just a part of the install routine and the only supported trigger method.

Load Sensing

The engine speed is one variable the ECU needs to determine how much fuel to add. But rpm by itself is not enough to make this calculation. An ECU that only used rpm would add the same amount of fuel if you were running flat out or had just lifted off the accelerator. The computer also

Looking down the outlet of a vane airflow meter, you can see the flap. This flap moves as the airflow through the meter increases.

This mass airflow sensor from JET has a hot-wire element located in the small tube at the top of the main tube.

needs to know how much *load* is placed on the engine. Most ECUs have what is called a *3D fuel map*. Two of the dimensions are rpm and a sensor used as a proxy for load, and the third dimension is how much fuel to add. However, they don't always measure load in the same way. While some ECUs measure how much air is coming into the engine directly, you can also get EFI running with less direct measures of load.

Alpha-N—One way to measure how hard you are pushing the engine would be to put a sensor on the throttle. The best sort of sensor to use here is a variable resistor (a potentiometer in tech speak, a TPS in car speak), and it usually sends the ECU more voltage the further the throttle opens. Using this throttle position sensor (TPS), as the main load sensor is known as *Alpha-N fuel metering*. It gets this name because *alpha* is used to mean the throttle angle and *N* is used to mean the number of times the engine fires its cylinders in a given amount of time.

Alpha-N is a pretty simple idea to understand, but not the easiest thing to tune, at least not for smooth drivability. At low rpm, you can often open the throttle very little and the engine will still be able to suck in nearly as much air as it would at full throttle, so tuning based purely on throttle opening can be a bit touchy. It does have it's place though

and is supported by many aftermarket EFI systems. Turbocharged engines can't be tuned on pure Alpha-N at all, as the amount of air getting by the throttle depends on both the throttle opening angle and how much the turbo has spooled up.

The throttle position sensor does have its advantages, however. On engines where the flow through the manifold becomes so turbulent or complex that it can confuse a sensor that measures the air, measuring the throttle gives you a good, steady reading. Alpha-N is popular on naturally aspirated racing engines with wild cams, and also is often used with independent throttle bodies. The TPS also is often the first sensor to know when the throttle has been opened. So even ECUs that do not use Alpha-N for their main fuel calculations frequently use it to add more fuel when you suddenly open the throttle (called *acceleration enrichment*, or *tip-in enrichment*).

A few engines, instead of using a TPS that returns a different voltage for every throttle position, have on/off switches. They'll tell the stock ECU that the throttle is closed, and maybe report if it's wide open. Such a TPS has very limited usefulness as far as aftermarket EFI is concerned and certainly couldn't be used to measure load. You can either just ignore this switch and measure load and tip-in differently, or replace it with a variable TPS if you need to use the TPS for these tasks.

Mass Airflow—The complete opposite approach from measuring throttle position is to directly measure how much air is flowing into the engine. There are several sensors that can measure this. One of the earliest designs is the vane airflow meter. This uses a flap that a spring holds closed, and the airflow pushes the flap open. A sensor (usually a potentiometer) measures the movement of the flap. This works, but it's a bit of a restriction to the airflow. Restriction is the sworn enemy of horsepower, so many people converting to aftermarket EFI like to yank out this kind of airflow meter and punt it like a football. (They are a bit heavy, so if you try actually punting it, it's a good idea to wear sturdy shoes.)

A more modern design is the hot wire mass airflow meter, or MAF for short. This holds a small heated wire in the airstream. The sensor keeps the wire at the same temperature as the incoming air tries to cool it down. The more the air flows past, the more current needs to flow through the hot wire to hold its temperature. The sensor sends a voltage or frequency based signal to the ECU to tell it how much air is coming into the engine. Since it only needs a wire sticking into the airstream to measure airflow, these can create very little

The familiar GM MAP sensor. This particular one is a three-bar sensor that works on engines running up to 29 psi of boost, which is the highest-rated pressure of a factory installed GM sensor. Aftermarket MAP sensors are available for even higher pressures.

High-powered turbo motors like Mark Wolfe's six-second Thunderbird experience a very wide range of airflow from idle to full throttle. Speed density is a logical choice for a motor like this. Photo courtesy Accel DFI/Prestolite Performance.

restriction if they're well designed. Though not all are well designed, and some MAF designs still cause a fairly significant airflow restriction.

Occasionally, you'll see a Karman vortex-type airflow meter. These create turbulence in the intake and use sound waves to measure the turbulence. These are complicated, rare, and restrictive. It's very rare to see standalone aftermarket ECUs that work with these. These also punt fairly well and can be easily replaced with a more suitable load measurement sensor for use with your EMS.

When installing a mass airflow sensor, you need to locate it in an area where it will get smooth airflow. Turbulence can throw off its reading. Also, be careful installing a mass airflow sensor downstream from a turbocharger or supercharger. Some will accurately read pressurized air; others won't. Factory turbo cars with vane airflow meters put them upstream of the turbo for a reason; it's difficult to get an accurate reading out of this sort of sensor under pressure.

A mass airflow meter sounds like it would be the ultimate way to measure airflow—it knows how much air the engine is getting, no matter what changes you've made to it. And they're very popular as original equipment. But in the performance world, they have a couple of limitations. When connecting one to an aftermarket ECU, you need to figure out how much voltage corresponds to how much air for every mass airflow sensor you hook it up to. Since the original manufacturer either built the meter themselves or worked closely with the designers, this is not a problem for them, but it's often hard for a backyard enthusiast to dig up these specifications.

The other potential issue is that the mass airflow

sensor must be sized to the engine. A MAF that has too large a range might not send an accurate enough signal when the engine is idling. Make the meter too small, and you can have worse problems: The meter will hit a point where it will stop telling the ECU it is getting any more air. A vane airflow meter can get its flap pinned all the way open, while a hot wire MAF will simply reach its maximum output voltage. If you're coming from the world of carburetors, this problem may sound a bit familiar. A mass airflow sensor can often cope with a wider range of airflow than a carburetor, but it still has limits. That said, a properly sized low restriction hot-wire MAF, properly calibrated to your EMS, is an excellent choice and possibly the most accurate air measurement device commonly used for EFI. For many this is challenging or expensive to get right, though, and there are easier and more affordable options.

Manifold Absolute Pressure Sensor—The last type of load sensor commonly used is the manifold absolute pressure, or MAP, sensor. This is the most commonly used load sensor with aftermarket ECUs. You can simply break down the name to determine what this sensor does and where it should take its reading from. *Manifold absolute pressure*—it reads the absolute pressure from the intake manifold. The pressure drops under a closed throttle and reads vacuum, approaches the same pressure as the outside air under full throttle (barring any intake plumbing restrictions), and rises above atmospheric pressure under boost on forced induction engines. An engine using a MAP sensor to determine fuel is called a *speed density system*, because it works off engine speed and the density of the incoming air.

Picking the right MAP sensor is often easier than selecting a junkyard MAF or finding one large enough for a very high-powered engine. They're typically sized in bar, where one bar is more or less equivalent to atmospheric pressure. So a naturally

Independent throttle bodies can make for a rough signal to the MAP sensor. Running a line from all the intake runners to a single small plenum smoothes out the signal and can make it easier to use speed density with such an intake.

A GM open element sensor for measuring air temperature.

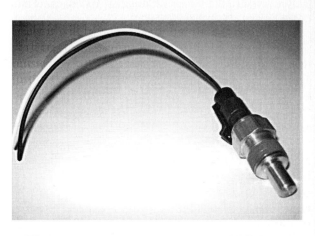

A GM closed element temperature sensor. While it can measure the temperature of either air or liquids, it doesn't react as quickly to rapidly changing temperatures like what you may see on a turbo car.

aspirated engine needs a one-bar MAP sensor minimum, but could use one rated for more. Engines running forced induction simply need one that can read more than the amount of boost they plan to run; for each 14.504 psi of boost, you need one more bar. This small number of sensor types is one reason speed density is very common in the aftermarket ECU world. Some systems even come with only one standard MAP sensor, since a sensor that reads four bar is still accurate enough to work on a naturally aspirated motor that never goes above one bar. A MAP sensor also works well on engines with very large changes in airflow, such as a high boost turbocharged engine that draws a minuscule amount of air idling but pulls in enormous amounts of air when the turbo spools up.

Speed density systems are not without their disadvantages though. One is that if you make significant modifications to the engine—particularly with the cylinder heads or camshaft—you will need to retune the ECU for best performance. They can also have trouble getting a good reading on engines that have large pressure waves in the intake, such as two stroke engines or race engines with very wild cams and minimal vacuum at idle. Independent throttle bodies can also make setting up a MAP sensor a bit trickier as the transition from a no-load reading to a full-load reading happens very quickly with ITBs with just the slightest blip of the throttle. A MAP sensor on such an engine needs to be connected to all the intake runners at once, and even then the reading may be a bit noisy.

Working Together—An engine is not restricted to just one sort of load sensor; it's possible, and even common, to design an ECU that uses two or even three different load sensors. The most common arrangement is to use a TPS for acceleration enrichment while using speed density or mass airflow for the main fuel calculations. Some ECUs have other ways of using different sensors

together. For example, you might have a turbocharged engine with a rumpity cam causing low vacuum at idle. The low vacuum at idle makes Alpha-N appealing, but Alpha-N can't handle boost, which will require a MAF or MAP sensor. One scheme used is to run a hybrid of Alpha-N and Speed Density, running alpha-N at low rpm, then transitioning to speed density as rpm or manifold pressure comes up and the MAP sensor can get a clearer signal. Or it may use an average of different sensor readings to calculate load.

Temperature Sensors

The engine's fuel needs change a lot with temperature. When the engine is cold, fuel has much more trouble vaporizing so the engine needs extra fuel to get a decent burn, which is why the engine needs more fuel when it's warming up—the colder it is, the more fuel it will need until it's warmer.

An intake air temperature sensor needs to be downstream of devices that add or subtract heat from the incoming air, but often gets a more accurate reading if it is in the intake plumbing instead of threaded into a metal, heat conducting manifold. This turbocharged Miata puts it between the intercooler and the throttle body.

Why You Can't Connect Two Devices to the Same Temperature Sensor

Sometimes it may seem tempting to connect two devices, like an ECU and a gauge or an aftermarket ECU and a stock ECU, to the same temperature sensor at the same time. Resist this temptation. While you can often get away with connecting two devices to a voltage-based sensor at the same time, resistance-based sensors don't work that way. An ECU or gauge determines the resistance of a sensor using a circuit called a *voltage divider*, which is a circuit that uses two (or sometimes more) resistors, one being the sensor and the others being inside the ECU. Connecting two devices to one sensor ties these voltage divider circuits together in ways the designers never intended. While this usually won't damage an ECU, the calibration in the ECU (and whatever else you have hooked up to the sensor) will be very far off.

When air temperature changes, so does its density and oxygen content. Before electronic fuel injection, carburetors used chokes to deal with a cold engine, and couldn't do much about air temperature changes except maybe warm up the incoming air with the exhaust pipes. With EFI, the ECU can simply measure the temperature of the engine and the air, and calculate the necessary adjustments.

Normal automotive temperature sensors are what in electrical-speak are called *negative temperature coefficient* (NTC) thermistors. This means that a temperature sensor is a resistor that has a high resistance when it's cold and a low resistance when it's hot. The ECU measures this resistance and uses it to determine the temperature.

Automotive temperature sensors come in three basic types. They are: open element sensors, which react very quickly to changes in air temperature but are not meant to be used in liquids; closed element sensors, which stand up well to liquids; and a special type of closed element sensor, used for measuring the temperature of metal parts. There are also thermocouples, but these are much more commonly used for exhaust gas temperature and will be covered later.

Coolant Temperature Sensor—The coolant temperature sensor is most commonly used for cold starting and adding more fuel while the engine warms up. Normally it's inside the waterjacket somewhere, upstream of the thermostat. On air-cooled engines, you can either measure the cylinder head temperature or the oil temperature to get the same function.

Intake Air Temperature Sensor—The *intake air temperature sensor* (IAT), sometimes called the *manifold air temperature* (MAT) or *intake charge temperature sensor*, helps the ECU calculate the density of the incoming air as a part of its calculations in determining how much fuel to add. Cold air is denser and has more oxygen, so the engine will need more fuel. Hot-wire type mass airflow sensors may have the temperature sensing built in, but all other types of load sensors will need an intake air temperature sensor. You can use either a closed or open element sensor for intake air temperature sensing, but an open element sensor will react faster and therefore provides a more timely and accurate reading.

Some parts of the intake system are better locations for the IAT sensor than others. If you put it upstream of a turbocharger, intercooler, or other parts that change the air temperature, it will not get the proper temperature reading to send on to the ECU as you want it to read the actual air temperature that's entering the engine, after these components have had their influence.

A more subtle problem can happen if you thread the sensor into the intake manifold, particularly on V-8 engines with an exhaust crossover in the intake manifold. In this case, the metal of the manifold can heat the sensor and give a false reading that's too hot, in a condition called *heat soak*. Some ECUs let you tune around this by using the coolant temperature sensor to correct the IAT sensor reading, but that still may not be the ideal. It's a difficult tradeoff, since ideally you would want to know the temperature of the air just at the point where the air enters the cylinders, but putting it in the manifold itself is likely to cause heat soak. The ideal location for the sensor is just before the

A Bosch LSU4.2 wideband oxygen sensor.

Since wideband oxygen sensors do not put out a straightforward voltage, it takes a special control circuit to interpret the data from them. The LC-1 from Innovate Motorsports is an example of a wideband controller, which translates the signal from the sensor into an analog voltage that it can send to an ECU or gauge.

throttle body as the air is entering the intake. The plumbing here should not cause severe heat soak issues, and you'll get actual air temperature as it enters the intake manifold. Note that this is an absolutely critical measurement for proper fuel calculations.

Sometimes turbo car owners run a second IAT sensor upstream of the intercooler and data log the information simply to check how well their intercooler is working, checking the temp before and after the intercooler does its job in various conditions. This is useful for checking in a data log to determine if you've got enough intercooler for the task at hand, but the ECU wouldn't need to use it for fuel calculations.

Exhaust Sensors

In the bad old days, only the government and the manufacturers had much in the way of equipment for examining a car's exhaust. While the sophisticated tools that are used in emissions tests are still a bit too expensive for most enthusiasts, these days

there are several tools that a do-it-yourselfer can use to examine the exhaust on a car and see how well the ECU is tuned. The most common sensors here are oxygen sensors and exhaust gas temperature (EGT) sensors.

O₂ Sensors—The most common type of oxygen sensor is the narrowband zirconia oxygen sensor. These are common enough that the part about zirconia is usually left off; the other type of narrowband sensor, the titania sensor, is very seldom seen and almost never paired with an aftermarket ECU. The zirconia sensor resembles a battery powered by exhaust gas, and puts out a small voltage. This voltage is around 0.1 to 0.2 volts when the mixture is leaner than a 14.7:1 ratio, and around 0.8 to 0.95 volts when the mixture is richer than a 14.7:1 ratio. There's a very steep slope to its output in the 14.7:1 area. The voltage it puts out in a rich mixture will depend on the exhaust gas temperature as much as the actual air to fuel ratio. This makes the sensor very accurate at pinpointing when the mixture is at 14.7:1, and horribly inaccurate anywhere else.

Most fuel-injected cars use narrowband sensors as original equipment because they're cheap, and they're good at telling the ECU when the car is at air/fuel ratio of 14.7:1, which is where it will be putting out minimum emissions. But they don't do a very good job of tracking your air/fuel ratio under full throttle. The difference between 13.8:1 and 11.8:1 AFR on gasoline could mean the difference between holes in the pistons and making safe power on a turbocharged engine running 20 psi of boost. But if your narrowband sensor reads 0.85 volts, you could be at either one. Most factory ECUs simply ignore the oxygen sensor at full throttle.

To solve this problem, engineers invented the wideband oxygen sensor. While this sensor design is related to the narrowband zirconia sensor, it is a much more complicated device, using a pump cell that works sort of like a battery in reverse, alongside a more conventional zirconia sensor. These also have heating elements to ensure they run at a controlled temperature, a feature that sometimes shows up on narrowband oxygen sensors too. The wideband oxygen sensor needs a controller to run the pump and interpret how the pump's activity relates to the reading from the sensor. This controller can be built into some ECUs, while most ECUs instead work with an external controller that is provided with the wideband sensor in a kit.

Inline engines often have no trouble running a single oxygen sensor. On a V-type engine or other motor with a true dual exhaust, running two wideband oxygen sensors can be expensive. If you'd

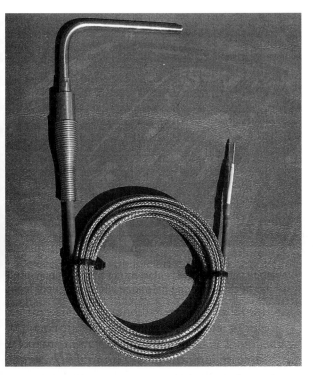

EGT probes are best at measuring cylinder-to-cylinder mixture variations. This small-block Chevy has one EGT probe in each header pipe.

Ordinary temperature sensors would not have the range to accurately measure exhaust gas temperature. Measuring this requires an EGT probe like this closed tipped probe.

rather not pay the expense, you can often get away with monitoring just one bank, or put your oxygen sensor in a crossover pipe between the two banks. Not all ECUs support using two oxygen sensors, but some of them can use two at once and correct each bank independently.

Oxygen sensors have several mounting requirements. Wideband sensors often need to be protected from the heat of turbochargers or otherwise high EGTs with the common Bosch LSU4.2 sensor having a max recommended temp of about 900 degrees F. They also should not be mounted where exhaust condensation will collect on the sensor, which typically means mounting the sensor with the tip pointing down (between 10 and 2 o'clock position) so condensation won't collect in the sensor tip. Also, it's vital to avoid any exhaust leaks upstream of the sensor, to prevent incoming oxygen from throwing the reading off. The sensor will usually come with instructions for where and how it can be installed. Virtually all oxygen sensors use the same 18 mm x 1.5 mm thread, so as long as heat is not an issue, you can simply put the wideband oxygen sensor where a narrowband sensor originally goes.

EGT Sensor—The exhaust gas temperature (EGT) sensor was a popular way to determine air/fuel ratio before affordable wideband sensors.

These use a type of temperature sensor called a thermocouple, which reacts quickly and is tough enough to stand up to the heat of the exhaust. As long as you're richer than the ideal stoichiometric air/fuel ratio (stoich) of 14.7:1, then the leaner you adjust the engine, the higher the exhaust gas temperature, up until peak EGTs at the stoich ratio. However, several other factors affect exhaust gas temperature, including ignition timing (retarding timing often makes the exhaust hotter). Since EGT probes are often cheaper than wideband oxygen sensors, some tuners use a single wideband oxygen sensor to measure the overall average AFR of all cylinders, and use an EGT probe at each cylinder to measure the cylinder-to-cylinder variations.

Barometric Pressure

Engines run differently in the thinner air at higher altitudes. On a carbureted engine, going to a different altitude meant rejetting the carburetor. With fuel injection, the ECU can use a sensor similar to a MAP sensor to measure the local air pressure. This can be a separate sensor (often inside the ECU), or a speed density system can use the MAP sensor reading on startup for the same function. Reading the MAP sensor at startup works well if the car is not going to deal with major altitude changes in a single trip, but getting the best performance at Pike's Peak demands an independent barometric pressure sensor to compensate for changes in altitude in real time.

Knock Sensing

Wouldn't it be great if you had a sensor that could do for your spark timing what the wideband oxygen sensor does for air/fuel mixture, and tell you just how much more or less you needed? Well, the closest things are dynos or high-speed cylinder pressure probes, neither of which are exactly cheap. There is a method called *ion sensing* that uses an ordinary spark plug to measure cylinder pressure, but so far this technology hasn't trickled down to aftermarket ECUs. So at the moment, the closest thing enthusiasts have is the knock sensor, which isn't really that close at all unfortunately, and isn't nearly as useful. While a knock sensor can tell when the tuning is very wrong, it can't tell when it is right. It can, however, notify a driver the he just filled up a tank of bad gas or that he has other unexpected problems causing knock on an otherwise well sorted tune.

The knock sensor is, quite literally, a microphone that the ECU uses to listen for the sound of the engine knocking. It can then adjust timing or fuel to reduce the knock, maybe even cut boost if the ECU is controlling a turbo. Knock sensors are not foolproof, as sometimes the ECU might not detect knock even when the driver can hear it, and at other times the ECU might mistake a noisy valvetrain or other engine noise for knock. OEM manufacturers often trash several engines in pre-production testing to get their ECUs to distinguish knock from other noises. And the sound characteristics of a motor could easily change if you've done something like replacing an iron cylinder head with an aluminum one. So while a knock sensor can be useful, it shouldn't be considered a substitute for careful tuning. It's more like the automotive equivalent of a smoke detector, a device that could possibly warn you of an impending disaster.

Fuel Sensors

The higher the fuel pressure, the more the injectors flow. A good fuel delivery system will maintain a constant pressure across the injectors, or change it in a predictable fashion. Keeping an eye on the fuel pressure, however, can warn you of potentially engine damaging problems with the fuel system. So some aftermarket ECUs offer a fuel pressure sensor as standard or optional equipment. The ECU could use a fuel pressure sensor simply for data logging, while it's also possible to make the ECU open the injectors longer if the pressure drops.

Fuel Composition Sensor—With the increasing popularity of flex-fuel cars, the ability to use a fuel composition sensor has started to show up in aftermarket ECUs as well as OEM ones. A fuel composition sensor can measure the ratio of alcohol to gasoline in the fuel, and is a key element to making a flex-fuel car practical to use. Ethanol demands considerably more fuel than gasoline for an equal amount of air. It's also higher octane, allowing more timing advance and, on turbo engines, more boost. An ECU with a fuel composition sensor can adjust the fuel and spark according to what's in your gas tank, letting you run any mixture of gasoline and ethanol that your fuel lines and injectors can tolerate.

Emissions Sensors and Other Devices

There are several other sensors that sometimes show up on fuel-injected engines, but most of them are there for emissions-control reasons. Many modern engines have sensors to tell if the catalytic converters are working, what the pressure in the EGR line is, and the like. These are seldom of concern on aftermarket ECUs, which are technically almost never legal to use on pollution-controlled vehicles anyway.

One type of sensor that can sometimes be useful with an aftermarket ECU is the kind that reports if the accessories are on. Many cars have switches to report when there is a large amount of electrical load on the system, the air conditioner turned on, or the power steering is working hard. Factory ECUs can use this to open the idle air control (IAC) valve a little further to kick up the idle and compensate for the extra load, and some aftermarket ECUs have general purpose inputs that can do the same sort of thing.

An all-out racing ignition setup on Gary Hart's 241 mph Studebaker.

Chapter 5

The Ignition System

While some of the first engine management systems used by vehicle manufacturers handled just fuel, most systems you'll encounter now can also manage the ignition. All but the most basic systems have a processor with enough power to handle ignition timing along with calculating the fuel requirements, and being able to adjust the spark along with the fuel allows you to tune your ignition advance curve on a dyno. And while you may occasionally see tuners who obsess about the last tenth of a point of air/fuel ratio, ignition tuning is where you will find the real power gains.

Many of the very earliest automobiles allowed the owner to tune the ignition in real time, in a manner of speaking. They had a lever or cable attached to the distributor that let the driver change the spark advance while driving. The driver took care of retarding the timing for cranking, advancing it for drivability or power. And since this was back when fuel varied tremendously in quality, this was also helpful for compensating for different grades of gasoline.

By the 1950s, you only had to adjust the timing as part of a tune-up. Designers removed the spark control from the inside of the car, and put a centrifugal mechanism and a vacuum can on the distributor that adjusted the timing automatically. Still, tuning this sort of thing represents a compromise, as there are only a few things you can adjust on a mechanical system—the weight of the flyweights, the strength of the springs, and the like. The centrifugal system limits you to a static rate of advance increase based on rpm only, and a set amount of max advance across the entire rpm

The reason the Studebaker needs a powerful spark. This photo shows a later configuration where the coil has been moved to the firewall and the MegaSquirt shown next to the coil in the previous panel is now controlling the timing, removing the need for a separate boost retard control.

range once you spin it fast enough to max it out. The vacuum advance adds some level of advance and retard based on load and gets around that to some degree but with nowhere near the flexibility of a computer-controlled programmable system.

An older GM HEI distributor controls ignition timing mechanically.

Looking inside the distributor, you can see a rotor that physically switches the spark from the coil to the terminals on the distributor cap, as well as the advance mechanism peaking out below the rotor.

A low-resolution (left) and a higher-resolution (right) trigger wheel.

Computer-controlled timing gives a far greater range of adjustment. Most systems that control the timing offer well over a hundred points of adjustment on the main timing map. Plus, a computer can adjust timing in ways a mechanical system could not handle—advancing the timing when the engine is cold, retarding timing if intake temperatures get too hot, or with a racing ECU, switching to an entirely different advance curve when you activate a nitrous system. They also give you the capability to run different amounts of ignition advance at different intake pressure (boost) levels, allowing you to crank up the boost and tune properly for it, without needing a separate little box-with-a-knob to handle that piece for you. It's all in the map. Done deal.

Engine Position Sensors Revisited

While you can inject the fuel at any time and have the engine still run, the timing of the spark must be pretty close to right for the engine to run at all, and needs to be spot-on for best performance. The same crankshaft position sensor or camshaft position sensor that sends the engine an rpm signal for the fuel injection typically also works to let the ECU know what position the engine is in so the ECU can calculate when to fire the spark.

Sensor Resolution—One issue spark control has that fuel control doesn't is the sensor's resolution.

As long as the total time an injector is open is right, exactly when the injector opens will only have a small effect on horsepower. With the ignition, however, timing is critical. If an engine only ran at a single rpm, resolution would not be an issue. In theory, you could have a single-toothed wheel on the cam for engine position on a V-8, and an ECU that could use this single trigger for all the spark timing calculations. In the real world, engines can change their speed abruptly, and the timing could have drifted quite far off before the tooth came back under the sensor and the ECU knew to update its calculations.

So the more points on the trigger wheel, the more often the ECU can update its calculations. However, too many trigger points can overwhelm the ECU's processing ability. An extreme example would be the ring of 360 slots found on certain Nissan distributors. The Nissan ECU used purpose-built hardware to count the slots. An ECU designed for a more conventional trigger wheel (which includes the majority of aftermarket standalone systems at the time of this writing) would try to fill in the gaps between the slots, in order to keep track of the engine's position in between signals from the sensor. An algorithm meant to do this would not be able to keep up with such a large number of slots at high rpm. It is far more common to either see one trigger point per cylinder, or a total of between 24 and 60 trigger points.

However, the resolution is not all that important if the trigger point passes under the sensor right when the spark plugs need to fire. Since the engine changes rpm most abruptly when cranking, it's common for designers to put the trigger points at the same place they need to be for the engine's cranking timing. An ECU programmed to take advantage of this can detect the appropriate trigger points and fire the plugs as soon as the tooth passes under the sensor while the engine is cranking. This strategy can be used on anything from a one tooth per cylinder setup to a 60-toothed crank wheel.

Older ignition systems moved either the trigger points or the sensor to change the timing as a function of rpm and vacuum. These systems used a set of flyweights, springs, and vacuum diaphragms to deliver a timing curve that, while not completely optimal, worked for its purposes. If you have one of these systems already dialed in well, you can choose to keep it in place and have the ECU control only the fuel. Since the sensor lines up with the trigger point automatically, there's no concern about resolution. A system like this might create problems for timing sequential injection, however. At the very least, you'd need an extra sensor for cylinder identification. For anyone new to engine management systems though, starting with a "fuel only conversion" and leaving the ignition as is will make your initial conversion much easier to take on. You're only worried about the fuel control for now, taking little bites off on this project at a time. You can come back and take over the ignition later if you want to, once you've got the fuel side working. You'll be more familiar with the computer, the software, and tuning principles by then. It gives you a chance to learn to walk before you run.

Even-Fire and Odd-Fire Engines

Most engines have the same number of crank degrees between cylinder firing. For example, on a conventional straight six, each cylinder fires 120 degrees of crank rotation after the next one. This is what's known as an *even-fire* engine.

Occasionally, you'll see what is known as an *odd-fire* engine, where there is a different number of crank degrees between different cylinders. For example, a 90-degree V-6 may need to fire one cylinder 90 degrees after the previous cylinder. It would then rotate another 150 degrees before firing another cylinder. Most odd-fire engines are V-type engines: Motorcycle V-twins and V-4s, 90-degree V-6s, and V-10s are often odd fire. There are a few odd-fire inline engines, such as the parallel twin design used on some Triumph motorcycles.

The uneven firing of an odd-fire engine makes them run less smoothly, so they have become less common in production cars. Designers can often avoid building an odd-fire engine with V-type designs by changing the geometry of the crankshaft so that two adjacent cylinders are not on the same

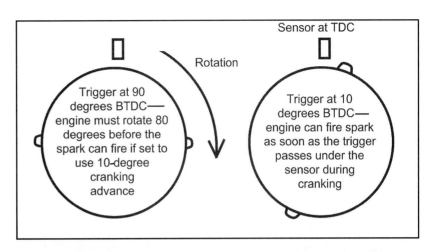

Two possible ways you could locate low-resolution trigger points on a four-cylinder engine with a distributor. The one with the first trigger point at 90 degrees BTDC is going to have a much harder time cranking than the one with the trigger point at 10 degrees BTDC.

If you have a 90-degree V-6, putting the connecting rods on the same journal gives you an odd-fire engine.

Making a 90-degree V-6 into an even-fire motor requires changing the crank design to separate the rod journals by 30 degrees. This can be done with a crank with six separate throws, or as shown here, with split-rod journals.

crank journal. If you're looking to control the ignition on an odd-fire engine, you'll need to check to make sure your ECU has this capability.

Ignition Coils

While under the right conditions, 12 volts shorted out could produce a spark large enough to start a fire, you're not going to see 12 volts jump a spark gap, and the pressure inside a cylinder makes the job even tougher. So ignition systems use one or more ignition coils to step up the voltage to something powerful enough to jump the gap at the spark plugs while under the high cylinders pressures of a running engine at full skip. This can be anywhere from 20,000 to 60,000 or more volts.

The ignition coil is a specialized type of electrical transformer. A transformer uses two coils of wire to change voltages. The low-voltage side is called the primary side of the coil, and typically has around

All modern ignition systems use a specialized kind of transformer called a coil to generate high enough voltages to fire the spark plugs. This performance coil is an example of the classic "can" style coil. Photo courtesy Pertronix.

The E-core coil is a newer type of coil that can be made smaller than a can-type coil of comparable power.

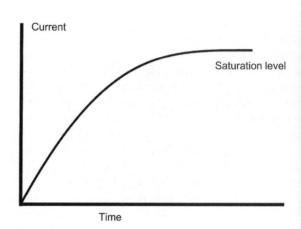

As a coil stores energy, the amount of current flowing through it increases. Eventually you will reach a point called coil saturation where the coil cannot store any more energy and the amount of current does not increase any more.

150 turns of wire wrapped around an iron core. The ECU or ignition module controls the current running through it. The other side is called the secondary, and it sends power to the spark plugs. The secondary coil often has tens of thousands of wire turns wrapped around the same core. A transformer can multiply the voltage by the ratio of the numbers of turns of wire when the current changes. For example, a coil with 150 primary turns and 30,000 secondary turns would multiply the voltage by 200. That would only let this coil step 12 volts up to 2,400 volts.

However, the coils don't just change voltage—they work like electromagnets, and can store energy in the magnetic field. Clever circuit designs take advantage of this to step up the voltage further, multiplying the coil output voltage by a factor of ten or more times than the turn ratio suggests you'd get.

The earliest ignition coils used a method called *inductive discharge*. To create an inductive discharge, you connect a 12-volt power source to one end of the primary wire, and ground the other end. Current will flow through the wire and build up a magnetic field. Then you disconnect the grounded end of the coil, causing the current to stop flowing. This makes the magnetic field collapse. The energy stored in this magnetic field creates a small voltage spike on the primary side. Since the transformer multiplies the voltage, this small voltage spike turns into an enormous voltage spike on the secondary side.

Modern inductive discharge ignitions use a power transistor like these Bosch BIP373s to control the flow of current through the coil. The transistor forms the core of an ignition module.

While running the current through the primary side of the coil builds up energy, there is a limit to how much energy the coil can store. As the coil is charging, the amount of current flowing through it will increase. The coil will eventually reach a saturation point where the current amount levels off, and it is not possible to put any more energy into it because the magnetic field cannot get any stronger. How long this takes will depend upon the coil. A canister-type coil from a 1950s car might take 15 milliseconds or more. Modern high-performance coils can take 2–4 milliseconds to fully charge. The amount of time required to charge the coil is known as *dwell time*.

Older inductive discharge ignitions use a mechanical switch called breaker points to turn the primary current on or off. Modern systems use a power transistor. The most common type of transistor used for this is the insulated gate bipolar

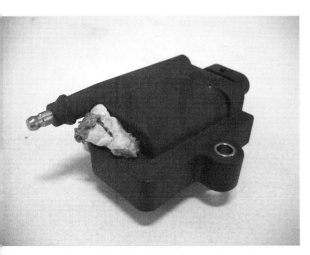

When you run a coil too long with too much dwell, the results aren't pretty. This coil is a very strong aftermarket coil, but the wrong ignition output settings caused it to run over ten minutes of dwell without discharging. If you're not sure how much dwell your coil needs, start at a conservative value.

transistor, or IGBT. These transistors can be built into an ECU, located in an external ignition module, or even built into the coil's housing itself.

Because a set of breaker points charges up the coil for a set amount of crank angle rather than a set amount of time, a coil designed for breaker points has to deal with a huge variety of dwell. If the driver turned on the ignition but did not start the car, you could even have the current flowing through the breaker points and the coil for minutes on end. A coil powerful enough to light off the spark at full throttle would draw a lot of current. Not only would a coil that could take this current continuously be very expensive, running that current through the coil for several minutes with the engine off would drain the battery, too.

Ballast Resistor—Designers solved this problem by adding a ballast resistor. This resistor is wired in series with the coil. As current flows through the ballast resistor, the resistor heats up and its resistance increases. The resistor cools off rapidly when the current stops. At high rpm, the resistance drops because the short dwell time gives it less time for heat to build up. At low rpm, the resistor heats up and limits the current. This saves the coil from overheating. Most points ignition systems bypass the ballast resistor during startup to deliver a more powerful spark. This does put more heat into the coil, but usually the cranking time is so short this will not damage anything.

Some early electronic ignition systems kept the ballast resistor. But it wasn't long before engineers figured out they could make an electronic ignition

Smart Modules and Dumb Modules

If you're using an external ignition module, you can divide these into two kinds: modules that control the dwell time, and modules that don't.

If the module does not control the dwell, the ECU needs to send it a pulse as long as the dwell. Depending on the module, it may start charging the coil when the ECU sends it a voltage and then fire it when the voltage stops. Or it may require the ECU to ground the wire connected to the ignition module to start charging, and fire the coil when that ground is removed and the voltage goes back up. When dealing with a module like this, you'll need to know which type of signal it needs and how long it charges the coil. Because they don't have any processing power or handle any calculations, they are sometimes known as *dumb modules*. These are the most common type of factory ignition module.

Other ignition modules contain a circuit that regulates the coil dwell time, so these can be called *smart modules*. Often the signal the ECU sends this module is a simple square wave, on half the time and off half the time. When wiring up one of these modules, you will need to know if it fires the coil when the voltage turns on or off. Capacitive discharge ignitions like the MSD-6A box always fall into this category because of their low dwell time, and CDI boxes frequently have a circuit to fire several sparks as well.

However, smart modules often aren't really that smart at all, as typically they simply follow a dwell time that is hard-wired into the module. Many of these modules do not measure the current or have a way of calculating what dwell time the coil needs. Consequently, a module with dwell control must be paired with a coil whose current draw and dwell time needs match the module's design. This could either be the same coil it came with from the factory, or an aftermarket performance coil designed to work with the module. While you can sometimes use a dumb module with a mismatched coil and change the dwell time with the ECU, taking a module that controls the dwell and pairing it with a coil that needs considerably less dwell time can fry the coil, the ignition module, or both. Sometimes "smart" modules know just enough to be dangerous.

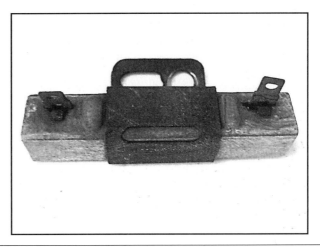

A ballast resistor from an old-fashioned breaker points ignition.

Calculating Dwell

How much dwell does a coil need? Give it too little and your coil will not put out as powerful a spark as it could. Give it too much and you can burn your coil or ignition module. If you're running a complete aftermarket performance ignition system including ignition module and coil(s), you can probably just call the manufacturer and ask them how much dwell you should be using. But if you're using stock coils or have mixed and matched systems, you may need another means of finding your dwell. Here are several choices depending on what equipment you have.

Hot Finger Method—If you don't have anything more sophisticated than a voltmeter, you may have to set the dwell by trial and error. Set the dwell low enough that the engine doesn't fire consistently and increase the dwell until it's around 0.2 milliseconds above the point where the engine misfires. If your coils or modules are known to be a bit on the fragile side, you'll need to keep a close eye, or rather a finger, on their temperature. If you find when tuning that you blow out spark under load, you may have to increase the dwell in small increments of 0.1–0.2ms. In most cases, the ignition module should not be warmer than a few degrees above room temp. If it is, you're probably running it to hard.

Oscilloscope Method—(Preferred if you have a scope) Use an oscilloscope to measure the current flowing into the coil. The current level rises as the coil charges. You should set the dwell so that it stays under the maximum current rating of your coil or ignition module (whichever is lower), or if it never reaches that, just shy of the amount of time it takes the current to level off or "current limit." This is where the steep slope of the current charging levels off after X number of amps. Once the module begins to limit the current things will just get hot from there; stop a bit short of current limiting and you've got that module doing all it can do with that coil. You'll also want to check the coil's specifications to make sure you aren't exceeding its maximum primary current.

Calculator Method—Measure the inductance of the coil's primary circuit or get this number from the coil manufacturer, and measure the resistance of the coil's primary circuit with a multimeter. Once you have these numbers, you can calculate the dwell required. You will need a scientific calculator with a natural logarithm (ln) function. Here is the equation:

dwell time = -(coil inductance ÷ coil resistance) x ln [1 – (coil resistance x maximum coil current) ÷ (maximum alternator voltage)]

The voltage is the highest you will reasonably expect the alternator voltage to reach, and you can often find the maximum current from the specs on your ignition module or coil. If the ignition module is built into the ECU, the ECU's manufacturer will usually be glad to provide the specs on the maximum recommended current, since this can save them a lot of time repairing an overheated unit.

limit current by building it so that it can only turn on for a short length of time before shutting itself back off. Since the ignition module then handles the job of limiting the maximum current, these ignitions do away with ballast resistors.

Dwell Limitations—The dwell time can limit the power and rpm that you can reach with an inductive discharge ignition. Using a single coil distributor, having a large cylinder count, and high rpm all constrain the amount of time the coil has to charge. A V-8 spinning at 6000 rpm requires 400 sparks each second. That works out to 2.5 milliseconds per spark. With most inductive coils that almost certainly will not be long enough to fully charge the coil, particularly since you have to allow up to a millisecond or so for the spark to burn before you can start charging the coil again. Realistically, this leaves only 1.5ms for the dwell time for each spark as you approach 6000 rpm on this engine. If the coil on this V-8 needed four milliseconds to reach saturation, its spark energy would start dropping long before redline. This is why a higher revving engine will benefit more from a coil upgrade than a slower turning motor will, and also why many convert to CDI ignition systems. A multi-coil (wasted spark or coil-on plug) can be a great solution to this problem as well. More on this in a moment.

Capacitive Discharge Ignition—The capacitive discharge ignition, or CDI, is one way to get around dwell limitations. The red MSD-6A box is probably the most famous example, but there are many similar designs on the market. Instead of storing energy in the coil's magnetic field, a capacitive discharge ignition stores energy in a charged capacitor. It then releases this energy in a high voltage pulse to the coil. The capacitors can charge quite rapidly, so these systems can deliver a full spark voltage at high rpm. CDI ignitions deliver a much quicker spark instead of one long spark. Most CDI ignitions fire multiple sparks at low rpm as a way of compensating for the short spark duration.

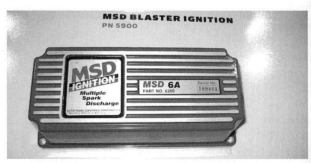

The MSD 6A is one of the most popular CDI ignition boxes of all time.

Not all CDI boxes are for distributors. This system from EMS is meant for coil-on-plug applications.

One other benefit of a CDI ignition is that it works better with engines that foul their plugs easily, whether it's from leaking oil or running rich at low rpm. An inductive ignition builds up its spark energy slowly, and if crud shorts out the gap on the spark plug, the energy can drain off and keep the plug from firing. A CDI ignition builds up the spark energy much faster, firing a short intense spark that is less likely to drain away through a carbon-covered electrode. Of course, it's better if you don't build up these deposits to begin with.

For a long time, the rule was that if you needed a hot ignition, you needed a CDI box. However, inductive discharge ignitions have become much more powerful in recent years. One of the biggest factors that has made this possible is the distributor-less ignition. Instead of having one coil for all the cylinders, distributorless ignitions use several coils. Having multiple coils allows each coil to charge for a longer period of time as it's no longer responsible for all eight cylinders—just one or two. So it has much more time to adequately charge before the next spark event. Using several coils also reduces the heat buildup in the coil, allowing coil designers to change their coils to handle more power with less worry that they will overheat. Some of today's hotter inductive discharge ignitions are more powerful than traditional capacitive discharge ignitions.

Distributor-Based Ignitions

The distributor is a type of rotary switch that allows one coil to fire all of the cylinders in an engine. The distributor simply uses a long strip of metal mounted on a rotor that drags the metal tip of this rotor past terminals inside its cap as it spins. The high-voltage spark jumps from the rotor to each of the terminals as it spins past, and then runs from each terminal down the spark plug wires to

the cylinder. Distributor-based systems are relatively inexpensive to build and do not require the ECU to be capable of identifying the cylinders. So they can be relatively straightforward to set up. However, because there is only one coil for all the cylinders, the distributor puts a limit on coil charge time at high rpm as we've already discussed.

If you are looking to get more spark energy out of a distributor, you can use a coil that charges faster requiring a shorter dwell time, or you can use a CDI system. However, the billet distributors that are popular in carbureted applications aren't as useful if you have a computer controlling the spark timing. Aftermarket distributors often are more precise and have easier-to-adjust advance mechanisms, but when there is no advance mechanism needed, the stock distributor won't hold you back as much as the stock distributors that come on carbureted engines might. On the other hand, if you don't have a distributor from a computer-controlled version of your engine, many billet distributors make it easy for you to lock out the advance mechanism without resorting to a welding torch so they can still make life a bit easier.

Double Distributor System—One peculiar variant on the distributor-based ignition is the double distributor. This shows up on some V-8 and V-12 engines. Two distributors, each sitting on one cylinder head, distribute the sparks from two coils to the spark plugs. This doubles the available dwell time compared to a single coil. However, it requires a computer almost sophisticated enough to run a distributorless ignition, all the sensors a distributorless ignition requires, and the mechanical complexity, extra spark plug wires, and extra wear points that distributor-based ignitions require. Some Bosch designs took this complexity even further, using two separate ECUs, each one of them controlling a separate distributor. This system did not stay around long, but you occasionally see it on Porsches, BMWs, and Lexuses. If you run into a dual-ECU version of this system, you can often

Two distributors mean half the dwell time but far more complexity.

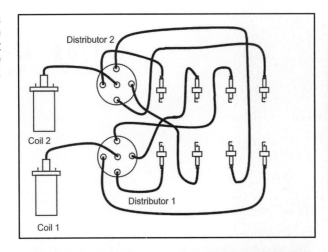

This Chrysler Hemi has dual plug heads and coil-on-plug ignition. Each coil attaches to two separate spark plugs, firing both at once.

Coil-per-plug systems do not have to stick the coils directly to the plugs. GM's LSx series uses valve cover mounted coils with short wires attaching them to the plugs. These coils are both quite powerful and easy to find in junkyards, making them a popular transplant onto other engines.

replace the two stock ECUs with a single aftermarket ECU.

Distributorless Ignition

When the ECU can identify the cylinders, it doesn't need a distributor to route the spark to the cylinders. Instead, the ECU can trigger separate coils for separate cylinders. Many of the earliest distributorless ignitions used what's known as the wasted spark principle. The coils in a wasted spark ignition fire two spark plugs at once, letting you build a distributorless ignition system with half as many coils as you have cylinders. The coil fires one cylinder on the compression stroke and the opposite cylinder on its exhaust stroke. Since the spark in the exhaust stroke neither has much effect nor drains much power, it's called a wasted spark. On a four-cylinder engine, a wasted spark ignition would give you twice as much time to charge the coils as a distributor. On a V-8, it has four times as much time compared to a conventional distributor.

Coil-on-Plug—The other type of distributorless ignition is coil-on-plug (COP). As its name implies, a coil-on-plug ignition has a coil attached to each spark plug. A variation called *coil-near-plug* has the coil connected to the spark plug with a short length of wire. Having one coil for each spark plug ensures that each coil has the longest amount of time to

charge as possible, as long as the coils are fired only on the compression stroke. Sometimes, you may need to set the coil to fire on both the compression and exhaust stroke, with the coils wired in pairs like they're in a wasted spark ignition. Usually you would do this if either you have more coils than your ECU has spark outputs, or your engine lacks a camshaft position sensor. Sometimes you'll hear people use the term "sequential COP" for true sequential coil-on-plug systems and "wasted spark COP" for systems using COP coils, but firing them in a wasted spark manner.

Sometimes you may be tempted to use a hot coil from a distributorless ignition in a distributor application. Although this may physically fit, you'll have to check the specs of the coil you have in mind to determine if it's a good idea. A coil in a distributorless ignition has much more time to cool down in between sparks. When pairing a coil intended for coil-near-plug use with a distributor, you'll find the coil firing much more frequently than it may have been designed for and not spending as much time cooling. It can run hot, reducing its spark output or even melting itself into a smelly glob of goo. Distributors generally require physically larger, better-cooled coils. The reverse scenario, running coils meant for a distributor in a coil-near-plug installation, isn't likely to be any problem for the coils. However, the sheer size of the coils can make this a bit awkward if you're using a set of traditional "can" coils for a coil-on-plug installation. Where are you going to put eight of those things? There are better options.

Electric fans are one of the more common things for an ECU to control outside of managing the air and fuel.

There's more to engine management than tuning fuel and ignition, even if that is its primary purpose. With a large sensor array and a lot of processing power, the ECU is a natural device for gathering data about your car and logging that data for you to review, controlling electrical devices, and taking the place of various other add-on electronics. You can program many aftermarket ECUs to limit the maximum rpm, act as a two-step or an electronic boost controller, or perform many other functions that traditionally you would have needed a separate piece of equipment to control.

Extra Outputs and Inputs

Extra outputs can let an ECU take the place of many types of single-purpose black boxes. This can simplify wiring, give you a single point to control everything, and even bring down costs compared to running a collection of rpm or temperature activated switches. This can simplify design whether you're a factory engineer or installing a standalone ECU. Here are some of the more common extra outputs.

On/Off—The most common function is a generic programmable on/off output. The idea is pretty simple. The ECU is already tracking a ton of information about your engine. Everything from coolant and intake temps to rpm to intake manifold pressure and more. So why not use this information to activate/deactivate accessories that need to be turned on/off based on information that's already available to the computer? You generally wire one of the ECU's generic outputs to control a relay, and wire that relay to

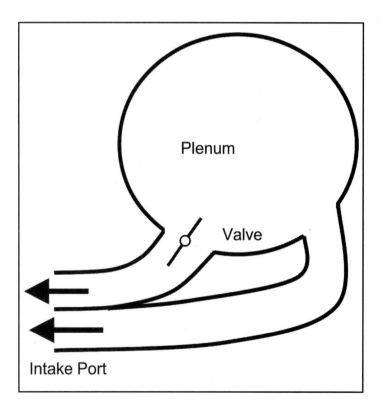

Hysteresis is a useful feature to prevent a device from cycling on and off too quickly. The fan turns on above 190 degrees and stays on until the temperature drops below 180.

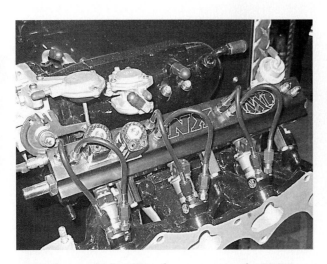

Many aftermarket standalone ECUs can control a nitrous system like this one, pulling timing and adding fuel when the nitrous is active. The ECU can take the place of an rpm-activated window switch and other nitrous control devices.

control your device. You then program the ECU using the tuning software to monitor whatever parameter you choose—coolant temperature for example, for when it gets over 180 degrees F. And when it does, turn on the fans. That's it. And voila, when your coolant reaches 180 degrees F, your fans come on. No need for a separate switch or another means of control. The parameter that's being tracked and used as a trigger can be as simple as the coolant temperature in the example we just used or manifold pressure/boost or even a combination of conditions based on the ECU's own internal calculations. Setting up generic outputs that are determined by the combination of specific conditions is doing what's called "AND-ing" or "OR-ing" of parameters. This basically means you can tell the output to trigger on or off only if two or more parameters line up the way you want them to. For example "IF manifold pressure is above 7 psi and throttle position is above 90%, then turn on the output that activates the water injection." Pretty cool, eh?

These outputs often have a feature called *hysteresis*. This keeps the output from rapidly cycling off or on if the variable it triggers on is sitting right near the trigger point. For example, if a cooling fan turned on every time the coolant temperature went over 200 degrees and shut off every time the temperature went below 200, and the temperature were right at 200, the fan would rapidly flicker back on and off as electronic noise made the temperature reading go between 199.9 and 200.1. So you might set the output's hysteresis to 5 degrees so that the fan turned on at 200 degrees and off at 195 degrees.

Usually these outputs work through relays. Occasionally, you may be able to drive low current devices like indicator lights or small solenoids directly. High current draw devices like cooling fans will always need relays. Check with the

manufacturer's documentation before wiring anything up directly that may draw more current than the output is designed to support, possibly damaging your ECU.

Cooling fan control is the most common use for generic on/off outputs. However, these outputs can be put to all kinds of other uses. Many newer engines have valves in the intake that open above a certain rpm, for instance. Honda's original VTEC system also operates with a simple on/off output. Creative hot rodders can find all kinds of uses for generic outputs.

A basic shift light can simply use a generic on/off output like any other device that turns on or off. Some ECUs can drive sequential shift lights instead, which light up more lights or different colored lights as an engine approaches its shift point.

Pulse Width Modulated (PWM)—A somewhat more sophisticated category of generic output is the generic PWM output. This is a pulse width modulated signal that is programmed using a 3D map, usually one with rpm versus engine load like your fuel tables. A generic PWM output is used for driving various devices that need to have continuous adjustment rather than simply being on or off. While most continuously adjustable systems benefit from specialized control code, a generic PWM output can be helpful for situations the ECU's designers may not have anticipated.

Nitrous—Most other outputs are a bit more specialized. Consider nitrous control, one popular feature in standalone systems. Since few drivers want the ECU to trigger the nitrous without permission, nitrous control calls for an input from an arming switch. Then the nitrous control can use certain criteria (such as rpm and throttle position) to tell when to turn on the nitrous with an on/off output. At the same time, a nitrous system often switches to a different fuel and timing map to make sure the tune is right for the additional cylinder pressure. Even though the hardware for a single stage of nitrous is only one input and one output, nitrous control is a bit more complicated than a basic on/off switch.

Nitrous control can sometimes be even more complicated. It's possible for an ECU with the right capabilities to kick in a second stage of nitrous above a set rpm or vehicle speed. Some nitrous systems even use pulse width modulated solenoid drives to continuously adjust the flow of nitrous.

Table Switching—There are other uses for switching fuel tables and spark tables beyond nitrous control, so it's common to see ECUs with an on/off input that allows you to switch fuel and spark tables. Table switching can let you make one

Many ECUs can drive a tachometer directly. Many Miatas starting in 1995 have ECU-controlled tachs.

set of maps for pump gas and one set of maps for race gas, or even separate maps for gasoline and a totally different type of fuel like propane or CNG (compressed natural gas) if you want. While carburetor tuners sometimes swap jets to have one set for best power and one set for best fuel economy, it's generally not a good idea to try and create separate fuel maps for power and economy. A map tuned for power at high load can still be set to be safely lean at low load for economical cruising where the extra fuel wouldn't do you much good anyways. Conversely, trying to lean-out high load sections of the map for a "wide open throttle fuel economy map" is likely to lead to disaster. With EFI and a properly tuned fuel map, the switch for going from gas mileage to power is located under your right foot.

Flex-Fuel—Flex-fuel input carries the idea of different maps for different gas one-step further. This uses the input from a fuel composition sensor, a special sensor that detects what mixture of fuel you are currently running and reports this to the ECU, to adjust the fuel and ignition for the gasoline when that's what you're running, or E85 ethanol when you fill up with that, or any combination of the two. You can switch fuels just by going to the gas station and pumping in whatever is available, and the ECU can then adjust the map to allow for the fuels' different air/fuel ratio requirements and octane levels. There are currently aftermarket standalone ECUs that fully support this functionality.

Gauges—Back when every car had a distributor, controlling a tachometer was simple: The tachometer read the rpm off the ignition output, which was a single pulse each time a cylinder fired. Capacitive discharge ignitions did little to change

this, as these boxes came with a tach output terminal even if they fired multiple sparks. The worst case was you'd need a little tach adapter box to get the voltage right. New ignition systems have rewritten the rules of how a tach works, as there may be multiple coils and the ECU determines rpm from reading a tooth pattern on a crankshaft position sensor.

While some tachometers are designed to read rpm from the coils or ignition modules in a distributor-less ignition, many tachometers do not quite work in such a situation. While you can rig an adapter circuit to drive the tachometer, it's also possible to have the ECU control the tachometer. As factory tachometers have used a wide variety of input signals, not all tachometer control circuits work with all tachometers. Sometimes there's no getting around using a tach adapter.

Some of the newest cars have taken ECU-controlled gauges a step further. Rather than having the ECU control just the tachometer, they've either placed the ECU in control of all the gauges, or have the ECU feed information to a computer in the gauge cluster via a network. It's not very common to find aftermarket ECUs that can tap into the network or drive factory gauges. On the other hand, some aftermarket ECU vendors have also come out with their own gauges or displays that read off a matching ECU. And there are third party digital dashes that can communicate with many aftermarket ECUs.

Boost Control—With the number of aftermarket electronic boost controllers out there, it's no surprise that aftermarket ECUs often come with a boost control feature. Factory turbo cars frequently have boost control built into the ECU as well. Boost control systems use a solenoid valve (or more rarely, a stepper valve) to operate a bleed on the turbo's wastegate, changing the amount of exhaust available to power the turbine. The simplest type of

Boost control functions help you get the most out of mammoth turbos.

The wastegate is a valve that controls the amount of exhaust flowing through the turbine. The ECU controls boost pressure by controlling the pressure acting on the wastegate.

The ECU controls a solenoid valve directly, varying how far the valve can open to send a signal to the turbo wastegate. This solenoid was used as original equipment on some GM applications and showed up in many aftermarket systems until GM discontinued production.

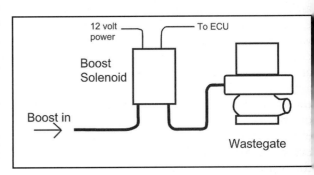

A diagram of how to plumb up a typical boost control solenoid.

boost control is open loop, where the amount the valve opens is a function of rpm and throttle position and you as the tuner adjust how hard to drive the boost valve (often the number is in duty cycle or DC) to increase boost. These systems work great and you can artfully craft a nice flat boost curve and then easily move the entire boost curve up or down according to your needs. If there's a weakness to this method, it's that there is no feedback loop. Since it's an open loop boost control method, the computer is not actually targeting a particular level of boost and adjusting the boost valve to get to that target; it's using static duty cycle numbers that the tuner programmed to drive the boost control valve. This isn't usually a big problem, but it can allow boost to drift up and down just a bit on its own, most often depending on weather conditions. Altitude and temperature are the biggest factors. A cold day can mean a bit more boost if you're running open loop control.

There are more sophisticated ways to control boost. Closed loop boost control causes the ECU to aim for a predetermined boost level, monitoring the boost level in real time and quickly adjusting the valve to maintain the proper targeted boost pressure

reading. Properly set up, these systems can help avoid any weather-related boost drift—unlike the open loop electronic boost control systems—as they control the wastegate to always provide the prescribed (tuned) boost level and to keep it there, regardless of conditions. These systems can take advantage of a tunable 3D map as well rather than simply a single boost level, or boost that adjusts only by rpm. They can allow you to tune for different levels of boost based on throttle position and rpm.

Depending on the system, boost control systems may also be programmed to reduce boost when the ECU senses knock, or reduce boost when the transmission is in lower gears to avoid sending the tires up in smoke.

Water Injection—Water injection is another extra often seen on boosted cars. Water cools the combustion chambers and acts as an anti-detonant. The water is sometimes mixed with methanol, most often in a 50/50 blend. This serves multiple purposes. The methanol still absorbs heat when it

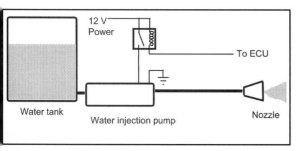

Wiring for a basic on/off water injection system.

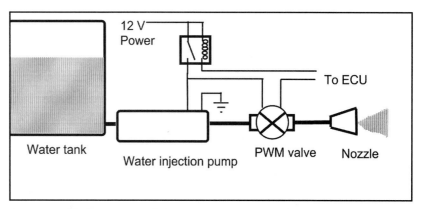

A more sophisticated water injection system using a fast acting valve to regulate a controlled amount of water into the engine.

vaporizes, it prevents the mixture from freezing if you're in conditions where this is a risk, and as a high octane fuel itself, it increases the octane of the overall air/fuel mixture, further reducing the chance of knock/detonation. A turbocharged car with water injection can get away with slightly leaner air/fuel ratios under boost than a comparable engine without water/meth injection since the water, instead of extra fuel, cools the combustion chamber. The water actually is much more efficient at cooling the cylinders than the extra fuel was anyway, and the air/fuel ratio can be kept closer to the ideal 12.5:1 in many cases. Also, more power can be made, the engine safely runs more fuel efficient under power, emissions improve, and everybody wins. A side benefit of water injection is the constant steam cleaning of your combustion chambers, blasting away carbon deposits and preventing new carbon from settling.

Water injection control may be a simple on/off solenoid output, spraying a high-pressure pump into a jet in your intake tract, spraying a constant static amount of water when activated. But many such systems use a PWM driver to control a fast-acting valve to vary the amount of water being added so it can be increased as the rpm and load increase and the engine is ingesting more air/fuel and therefore can ingest more water as well. Depending on the system, it may also be possible to have the ECU turn on the water pump, or pull timing and add fuel if sensors indicate you've run out of water. On ECUs that don't support the latter feature directly, you may be able to use a table switching input and a fluid level switch for the same result.

Variable Valve Timing—There are several types of variable valve timing systems out there, so if you're looking for an ECU to control a factory variable valve timing system, you will need to know how your stock system works. The earliest variable valve timing setups use a simple on/off switch. Some, like the early Honda VTEC systems, added an oil pressure switch that would detect if the oil

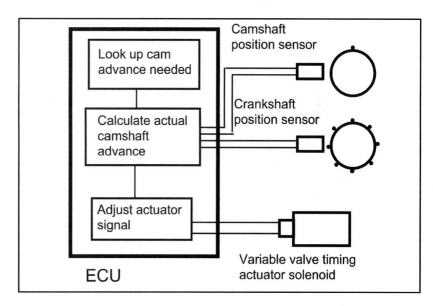

Closed loop variable valve timing uses input from the crankshaft and camshaft position sensors to make sure the cam advance is in the correct position.

pressure was too low to activate VTEC. These systems are pretty simple from both a hardware and a tuning perspective; you can set them to activate at a particular rpm, ideally when the engine's torque curves on the low-rpm and high-rpm lobes intersect so you get the most out of each cam profile.

Later variable valve timing systems allowed continuous adjustment of the valve timing, and sometimes separate control of the valve lift. These can take several forms, but the most common is a pulse width modulated solenoid that lets the ECU vary the oil pressure to an actuator. Some systems have used DC servomotors instead of, or even alongside, the PWM solenoids. Continuously variable valve timing makes tuning a bit more complicated. While you could get a crude measure of control with a generic PWM output, getting a

A drive-by-wire throttle body.

system to truly work is more complicated for a few reasons.

The first area of complexity is making sure the cams actually advance or retard the timing as much as they should. If the control solenoid opens a certain amount, how much the cam actually moves will depend on the oil pressure and other characteristics of the oil. So this calls for closed loop control, where the ECU can detect the amount of cam advance by comparing the camshaft position sensor signal to the signal from the crankshaft position sensor. Without closed loop feedback, changing the oil and putting a slightly different thickness in your crankcase could throw off your valve timing, and throwing off the valve timing will throw off the tuning.

The second area that continuously variable valve timing complicates is the tuning. The volumetric efficiency changes with the cam timing, and if both intake and exhaust cams change timing, it will also change depending on the lobe separation angle (remember anytime you move those cams, you've just created a new engine you need to tune for). Your ECU needs a way to determine how the change in cam angle affects the airflow and how to adjust the fuel and spark to compensate. Some factory ECUs contain multiple VE tables and switch between them, depending on the position of the cams. Other OEM systems use complex engine modeling techniques to create a computer model of how the engine will behave and predict the needs based on this model, eliminating the need for multiple tables.

One of the most common approaches in the aftermarket ECU field is to simplify control of the system and to move the cams only once or twice instead of constantly. Then tables can be tuned for the different positions and blended together. The second, more complex but technically superior approach, is typically creating a table that maps out the cam positioning under different load/rpm conditions, causing the cams to be in the same places every time when at a given load/rpm point, and then you just tune the engine like normal—no need for multiple tuning maps.

If you're looking to control variable valve timing, you'll need to find out precisely how it is done on your engine and compare it to the ECU's features. If your ECU of choice does not support this, you'll either have to lock out the cam timing or leave the stock ECU in place to control it, possibly in parallel with your aftermarket ECU controlling the rest of the show.

Computer-Controlled Throttle—Drive-by-wire is a feature that first surfaced in the late '80s on production cars, but has recently become much more common. Instead of mechanically connecting the throttle to the accelerator pedal, the ECU controls throttle opening with an electronic actuator. Part of this is to reduce the vibrations the driver feels, and it also can be used for idle speed control. Other possible applications can include traction control or lifting off the throttle while shifting an automatic transmission. Setting up one of these requires extreme caution. You don't want a sensor failure or a bad line of code to pin the throttle wide open. Often your best approach when swapping a drive-by-wire motor into an earlier car is to use a cable-operated throttle body.

Engine Control Tricks

While adjusting the fuel and spark tables deals with most of the engine's tuning, many aftermarket ECUs have additional tricks up their sleeves with the engine control. Normally you would want to tune an engine so it runs as smoothly and efficiently as possible. But many tricks revolve around deliberately reducing the engine's power output, or even making the engine run badly on purpose.

Rev Limiters and Relatives—You may be wondering why a tuner might want to sabotage an engine's ability to run well. The most obvious reason is the rev limiter, which would prevent the engine from revving smoothly through its powerband, past its redline, and sometimes right into the throw-the-rods-through-the-block point on the tach. There are several ways to create a rev limit, and many ECUs go through this in stages. Of course, no rev limiter can prevent certain types of overrevving, like downshifting when the engine is already at its redline.

Often the first step is to simply retard the spark timing to reduce power. It usually works well with the engine in gear, but often won't prevent the engine from revving up in neutral (or if something in the drivetrain breaks!). ECUs that lack a spark retard rev limiter can often have such a limiter programmed into the spark advance table by setting your last column at the highest rpm to a severely retarded timing figure, maybe in the range of 0–5 degrees. Much of the engine's power will drop out, but this still won't be a true limiter; it will just slow the train.

If the spark retard isn't enough, the next option is to prevent the cylinders from firing at all, using

Dial in a launch rev limiter correctly, and you'll be able to leave the starting line at full throttle without spinning the tires. Traction control takes this further, reducing power output if the wheels start to slip. Photo courtesy BigStuff3.

either a *fuel cut*, a *spark cut*, or both. Which one to use can depend on the engine and vehicle. Cutting the spark alone will dump raw fuel into the exhaust, which is one reason why nearly all factory rev limiters cut fuel instead. However, a fuel cut needs to be a complete fuel cut, while you can make a spark cut kick in gradually by dropping some of the sparks but not others. You can't partially cut out fuel, which is the same thing as running your engine lean, which is definitely not the goal. Fuel cuts are also not appropriate if you have extra sources of fuel, such as a wet-flow nitrous system. Cutting off all the fuel will avoid a lean situation and the resulting potential detonation because there's no burning, premature or otherwise. But if you let fuel in from a source other than the injectors, you're in serious trouble with a fuel cut as you could have just enough fuel for lean detonation to occur. So with a fuel cut, it's all or nothing.

There are other uses for a rev limiter besides keeping the engine from damaging itself. Many standalone ECUs offer rev limiters that kick in at other points. A launch rev limiter (often called a two-step) lets you set a lower rev limit at the dragstrip, hold the accelerator to the floor with the clutch in, and hold the engine at an rpm that you set. Once you release the clutch, the rev limiter turns off. A flat shift rev limiter is one that engages when you put the clutch in to shift, enabling you to shift without lifting off the gas pedal.

There are also a variety of anti-lag systems that are often close relatives of the rev limiter. These are designed to burn raw fuel in the exhaust to spool up a turbine. Sometimes you can use a spark cut rev limiter/flat shift algorithm to get a working antilag system. Other antilag systems are more

sophisticated, with injectors in the exhaust or other direct methods of ensuring that there's still a fire burning in the headers. You should know that antilag systems, while very cool in concept, are very rough on turbos and other components. Great fun if you don't mind replacing turbos, headers and exhaust valves on a regular basis, but not exactly wise to use on a car that needs to last long on a limited maintenance budget.

Many traction control algorithms are also in the same family. This type of traction control system retards the spark to control wheelspin. Some systems will also add fuel, creating an overrich condition, which will reduce power as well. Either way, the goal is to reduce torque when wheelspin occurs. More sophisticated systems can reduce the throttle, but this capability is exceedingly rare in standalone ECUs at this time. Some basic traction controlled systems are more accurately called *anti-rev* or *slew control systems*. They simply detect if the rpm is increasing too quickly. More sophisticated traction control systems use vehicle speed sensors on both the drive wheels and the non-driven wheels and compare readings from each to determine if the drive wheels are spinning.

Multiple Tables—Most commonly, all the injectors will be equally sized and will squirt the same amount of fuel, whether they're sequential or batch fired. However, if the ECU has more than one injector driver (either a sequential or a bank-to-bank system), this doesn't have to be the case. An ECU may have individual fuel trim settings for its injector outputs. This can be either a fixed percentage, or a table. This lets the ECU compensate for cylinder-to-cylinder differences in air/fuel ratio. Tuning this feature correctly can be an expensive and time-consuming task, but it can be useful to get every last bit of power out of a motor. In the case of a bank-to-bank system, you may be able to use separate tuning tables if you either have one unusual cylinder, or half the cylinders working well off one table and half working off another. For example, the big-block Chevy has a peculiar combination of equally spaced exhaust ports and book-fold intake ports, so half their intake ports tend to flow better than the other half.

Another case where you do not have all the injectors doing the same thing is *staged injection*. Staged injection typically has two injectors per cylinder, but other combinations are possible, such as one injector for each cylinder and an extra injector in the plenum. An engine with staged injection will drive around on one set of injectors when fuel demand is low, with the second set kicking in when fuel demands increase. This lets

Sometimes eight injectors just aren't enough. With staged injection, you could assign the upper row a separate fuel table from the lower row, and have the upper row only kick in when you need it.

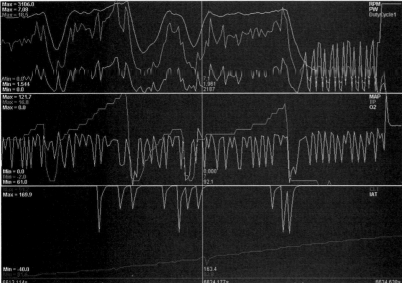

Data logging is a very handy tool for troubleshooting your engine. This data log came from an engine that was nearly impossible to tune. Sifting through the log revealed two main problems: The MAP sensor could not get a good signal, and the intake air temperature sensor was not working. An engine like this would run better with alpha-N, and the data log demonstrates why. The noisy MAP signal is causing the injector pulse width to become equally noisy.

you set a long, easily tuned pulse width at idle and low-speed driving using a smaller set of injectors, while allowing enough fuel delivery to make huge amounts of power when the second stage of injectors come in, often much larger. Staged injection usually turns up on turbo engines running large amounts of boost, or other engines that must accommodate a wide range of power output. Some ECUs that do staged injection make an abrupt transition, while others let you gradually increase the amount of fuel from the secondary injectors. You can even give the secondary injectors a separate fuel system loaded with a higher-octane fuel so you only burn the expensive stuff when you need it.

Data Logging

Data logging turns a standalone ECU into even more than just an engine controller. This ability lets you record the sensor readings and the ECU's outputs in a file you can play back and examine. Depending on its features, you can use data logs to do anything from analyze the quality of your tuning to examine the health of the engine to find ways to improve your lap times.

Types of Data Logging—There are two main types of data logging: external and internal. External data logging means that the ECU is not able to store the data log itself, but it can send the data to a laptop which then saves the file. Speaking from experience, sending a laptop down the Bonneville salt duct taped to the passenger seat at 200+mph works, but it's not ideal. ECUs supporting external data logging sometimes uses a different device, such as a Palm handheld or a dedicated logging device to save the file. These smaller devices can make data logging on the go, particularly in a racing environment, much more convenient. Internal data logging means that the ECU can record the data without connecting any other device to it, so you can retrieve the data after a race or test/tuning session. Sometimes it is possible to put marks in the data log by pressing a key or button if something happens during your test that you want to to be able to easily find in the log, letting you know what areas to examine later.

How to Use the Data—Once you have a data log, there are several things you can do with it. The most common use is checking your state of tune for further refinement if needed. This works best if you have a wideband oxygen sensor connected to the ECU. You can then view the oxygen sensor readings and compare them to the rpm and load to see if your air/fuel ratios are what you need them to be. Depending on your system, you may even be able to use the data logging software to edit your fuel tables based on the information in the data log.

Data logs can also diagnose engine problems. If you are having trouble at a particular rpm, you can examine conditions at that spot on the graph and see if any readings are out of the ordinary. For example, you may see that the engine is running lean, or that a noisy signal from the crankshaft position sensor at a particular spot is throwing your calculations out of whack. You can also use data logs to gauge how well the electrical system is working—a data log can reveal noise in various sensor signals or point to a drop in battery voltage under the wrong conditions.

Data logs can even be useful for analyzing how a car performs during a race. At the very least, a data

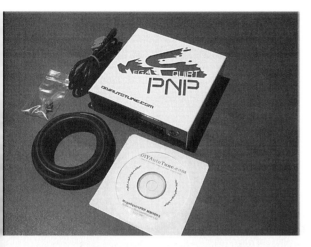

This plug-and-play MegaSquirt ECU directly replaces the factory ECU, with the same type of connector built in.

A look inside a Link ECU shows the engine management circuit board on top attached to a lower circuit board that adapts it to the stock ECU connector.

A plug-and-play system can also take the form of a harness with a stock ECU connector on one end and the aftermarket ECU's connector on the other. Photo courtesy Haltech.

log from a race will record when how long you spent at a particular rpm, what your coolant and air temperature readings were during the race, and maybe information from a wideband oxygen sensor letting you know your air/fuel ratio under various conditions. If you are able to record information from a GPS, accelerometer, or vehicle speed sensor in the data log, you can even use this to examine a car's position on a race track and possibly locate spots where the car can be driven harder—or was pushed beyond its limits.

"Plug-and-Play" Systems

If your car was originally fuel injected and the stock computer won't cut it anymore, you may be able to find a plug-and-play standalone. These computers, as their name implies, come wired up to plug into the factory wiring with little or no wiring or other changes required to get the hardware installed. This could be by putting a stock-type connector on the ECU itself, or in other cases by using an adapter harness included with the unit. Often a plug-and-play system will come with a base map for the car as a starting point, although it will need tuning to get the most out of the engine.

Not all plug-and-play systems are direct fits for the stock ECU using 100% of the factory sensors. Some of them require minor changes such as routing a MAP sensor hose or the like. A few plug-and-play systems on the market require some wire splicing to install, although they mostly use the factory wiring harness. And if a plug-and-play ECU has capabilities the factory ECU didn't originally have (that you want to take advantage of), you may need to run additional wires for new sensors or actuators. A plug-and-play ECU for a 5.0 Mustang may offer nitrous control, but it has

nothing in the stock engine compartment to control the nitrous with, so a feature like this would require new wiring.

Using Plug-and-Play with Engine Swaps—If you're going with a plug-and-play ECU, be careful with engine swaps. You can retune an aftermarket ECU for different camshafts or a different size engine, but if you end up using sensors that operate differently from the ones you're replacing or a different ignition system, your plug-and-play system may not work with the new engine. As a general rule, if the engine you're swapping in could be plugged into the factory ECU and run, even if it runs badly, you can probably use a plug-and-play ECU for your car. You'll want to check with the ECU vendor if you're planning any sort of drastic engine swap.

Sometimes there's another option for engine swappers: preassembled harnesses. These harnesses can save the time of fabricating a new harness. They are particularly common in the domestic V-8 world, where owners often upgrade classic muscle cars with late-model, fuel-injected motors. Usually these harnesses go with the engine, not the chassis, so they will typically include just the engine control wiring designed to work with a specific ECU.

Chapter 7
Factory Stock ECUs

The vacuum canister on this distributor is a sure sign the ECU is not controlling the timing. At most, you might sometimes see a mechanical distributor advance paired with a knock retard system.

Now that we've gone over the building blocks of an EFI system, we'll take a look at how the factory does things. This information can come in handy whether you want to remove, reprogram, or fool the stock ECU. Many aftermarket systems will let you calibrate the ECU to match all kinds of factory sensors, although some of the standalone systems aren't so flexible and will always require a specific set of sensors that are usually offered along with the system. And some factory sensors or trigger wheels are so obscure they pretty much require an ECU designed explicitly for them. If you're going with a standalone system on one of these, no matter what standalone system you're going with, you will be fitting some new elements to the system. Basically, you'll need to do a bit of homework to determine what your engine uses, and what your ECU of choice will support. This may even affect your choice of ECU, leading you to choose a model that will natively support your factory hardware, if available. That's a bit vague; let's discuss some specifics.

First off, let's eliminate what you really don't need to be that concerned with. Some electronic components of an EFI system are virtually the same for almost any system. Except for a few very early designs, most temperature sensors work according to the same principles even if they don't have the same calibration curve. Most standalone systems support speed density using a MAP sensor, so you won't need to figure out how to connect them to a mass airflow system unless you really want to. And most factory oxygen sensors behave in the same way unless you have a vehicle with a wideband sensor from the factory (somewhat rare, but becoming more common in recent years). Throttle position sensors are generally either switch (on-off) type, which only tells you if the throttle is wide open or fully closed, or variable (potentiometer) type, which are able to provide an accurate throttle position across the entire range of motion. These are the common similarities you'll find among nearly all EFI systems from the factory, and the things that can be more easily taken for granted when planning out your system because they're pretty similar from engine to engine. This chapter will, for the most part, focus on how the various OEM fuel injection systems differ from each other, and the pieces you will need to make sure your conversion is successful. We will also touch on some of the tuning options for factory ECUs.

Injectors usually only differ by their resistance unless you have a newer vehicle with direct injection into the combustion chamber or something along the lines of a Bosch K-Jetronic (which isn't EFI at all, but mechanical fuel injection). So for injectors, you simply need to make sure your choice of system can control your injectors. High-impedance injectors should never be a problem, though low-impedance injectors put more load on the injector drivers and the ECU needs to be built to handle this. Make sure your choice of ECU is ready for the task if you're using low-impedance injectors. There are ways to control low-z injectors with an ECU intended to control hi-z injectors if you need to, however. See Chapter 2 for more information on this if you missed it on the first pass.

Factory ECU Differences in Operation

The biggest difference between factory systems is often how they handle the ignition control, or breaking that into its two pieces, how the factory system determines engine position and rpm, and then how it goes about firing the spark plugs. Those are the two questions you should be trying to answer early on. How the spark plugs are fired is the more obvious of the two. The designers may use a single coil with a distributor or multiple coils. The coil drivers may be inside the ECU, inside the coil assembly, or in another device (commonly called an igniter) between the ECU and the coil.

The second piece of that puzzle is that engineers have come up with many different ways to tell the ECU where the engine position is. Let's split this up into a few different categories, starting with the old school and moving on into modern times.

Mechanical Timing Controls—The first category is mechanical timing controls. On a car like this, the ECU does not handle ignition control at all, or at the most, simply transmits a signal to the ignition module to retard the timing when it detects knock. A mechanical device inside the distributor determines the ignition timing, and the ECU just gets a tach pulse each time the cylinder fires. This is your old school mechanical/vacuum advance distributor. The only additional purpose it may be serving on an early EFI car is to give the EFI computer an rpm signal from an internal trigger wheel so that the EFI computer can calculate the fueling needs based on this signal and its other inputs. This would be an example of a fuel-only EFI system as that's all the computer is managing, the fuel.

Distributor System—The second category is what I will call a basic distributor system. This still has a distributor but without any means of mechanical advance; instead the ECU is completely in charge of the spark timing. A pickup, often inside the base of the distributor, sends the ECU a single pulse each time a cylinder reaches a particular crank angle before top dead center, and the ECU then starts a countdown until it needs to fire the single coil. These systems may have a distributor with a trigger wheel containing one tooth for each

cylinder, or a crank trigger with half as many teeth as the engine has cylinders. An engine that originally had mechanical timing controls could be converted to this category if you locked out the distributor's advance mechanism when converting it to computer control.

Dual-Wheel System—The third category is a basic dual-wheel system. This has a primary pickup reading a wheel with equally spaced triggers, like the basic distributor system. But there is also a secondary pickup that sends the ECU a sync pulse once every cam revolution, or more rarely, once every crank revolution. The primary pickup may send one pulse per ignition event, in which the secondary trigger only provides cylinder identification for sequential EFI or for a distributorless ignition (or both). Such a dual-wheel system could be used without the secondary as a pickup for a basic distributor system in many cases. Other times, the primary pickup will send several pulses per ignition event (such as a 24-toothed wheel on a six cylinder, providing four teeth per ignition event), in which case you absolutely must use the second sensor to identify the cylinders, even if you have a distributor. To confuse matters, there is sometimes a third pickup, also reading a wheel with equally spaced teeth, but with fewer teeth than the primary wheel. OBD1 and later Honda distributors, for instance, have three wheels: a 24-tooth, a one-tooth, and a four-tooth. Since it's a distributor-based ignition system, you've really got all the information you need with that four-tooth wheel. But using the 24 and the one-tooth wheel together can provide more accurate crank position information and allow for other interesting options, like adapting your ECU to fire a sequential COP

Different examples of trigger wheel patterns: dual wheel with no missing teeth, m-n, and irregular teeth spacing.

VW was the first company to use Bosch EFI in the 1960s, and they've used engine management built by or licensed from Bosch ever since.

ignition system rather than the stock single coil system. Basically you'd be using the distributor as a CAS but nothing more.

M-N Wheel—A fourth category is the m-n wheel. This has a wheel that would have m equally spaced teeth on it, except there is a gap of n missing teeth. For example, a 60–2 (as in "sixty minus two") wheel would have 58 teeth and a gap two teeth apart. The wheel usually spins at crank speed, but more rarely you may see one spun at cam speed. The ECU uses the gap for cylinder identification. Often you'll see an m-n crank wheel paired with a one-tooth cam wheel, like in the basic dual-wheel system except the teeth aren't all equally spaced. These setups can be used with either a distributor or a distributorless ignition system. You can commonly buy 60–2 or 36–1 reluctor wheels for use with aftermarket engine management systems to fit onto engines that used other trigger patterns that your EMS may not be able to decode, allowing you to use pretty much any EMS with pretty much any engine if you fit a compatible trigger wheel.

Other Wheels—The fifth category is a catch-all term for unusual trigger patterns that frequently require specialized programming to read. For various reasons, manufacturers have come up with several different trigger wheel patterns that use irregular spacing of the trigger wheel teeth. Sometimes this is as simple as equally spaced teeth with one tooth added between the teeth (We'll call this one an m+1 wheel). Or a trigger wheel may have several small clusters of teeth separated by large gaps. These usually require code specifically written to deal with the wheel pattern, although sometimes you can grind a couple teeth off such a wheel to make it fall into one of the previous categories. If your wheel does not allow this, and the designers of your ECU of choice did not give it built-in support for your trigger wheel, you may need to retrofit your engine with an aftermarket crank trigger.

Bosch EFI systems

Bosch didn't invent electronic fuel injection, or build the first production automotive EFI system. Those honors go to Bendix, but the first Bendix system, the Electrojector, was a complete disaster when it hit the market in 1958. After the Electrojector, Bosch bought part of Bendix's technology and refined it into the first practical EFI system, the D-Jetronic. Bosch is still one of the biggest EFI suppliers worldwide. Their complete engine management systems, or direct copies, appear in most BMW, Mercedes, and Volkswagen cars. Their systems have also been used in Nissans, Fiats, Yugos, Hyundais, Kias, and many European cars not sold in the United States. Even companies that use ECUs from other sources often use Bosch components.

D-Jetronic EFI—The D-Jetronic was Bosch's first system. It was a speed density analog system, the D standing for *Druck*, German for "pressure." The D-Jetronic used mechanical timing controls with a breaker-point distributor and a second set of points in the distributor that created a tach pulse to trigger the fuel injection. D-Jetronic systems used an unusual throttle position sensor that generated pulses as the throttle opened, instead of a voltage that rose with increasing throttle.

Note: This means the D-Jet TPS is incompatible with many aftermarket ECUs, though you can often fit a compatible TPS in its place, or run without one. Many of its other sensors were equally unusual by modern standards, such as using a variable transformer for a MAP sensor.

K-Jetronic EFI—Bosch also built a system called the K-Jetronic or CIS, which started out as a mechanical fuel injection system that later acquired some help from electronic controls. Very few aftermarket standalone systems can control the electronic pressure controls on these systems. If you are putting aftermarket EFI on a car with a K-Jetronic-type system, you will need to replace the injectors and the fuel pressure regulator with parts designed for EFI.

L-Jetronic EFI—Bosch's modern EFI systems started with the L-Jetronic, a digital system that used a mass airflow sensor. The L stands for *Luft*, German for "air." The earliest versions used a vane airflow meter, while the later LH-Jetronic used a hot wire mass airflow sensor. The L-Jetronic did not control timing, but instead of mechanical timing controls, many of the L-Jetronic systems used separate electronic timing control devices, which grew more elaborate the longer the system stayed in production. L-Jetronic systems typically have a basic distributor setup with a Hall effect sensor in

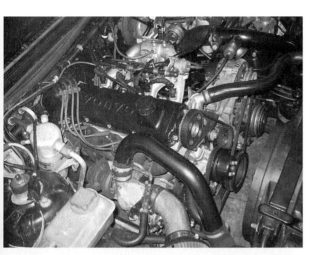

This Volvo 940 originally used a Bosch L-Jetronic system to control the fuel and a separate control module to control the timing.

The 5.0 Mustang is one of the most frequently tuned cars to come with an EEC-IV ECU.

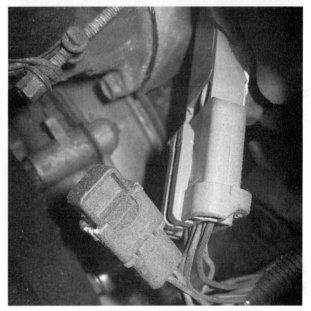

The SPOUT diagnostic connector on a TFI ignition. Pulling this connector locks out the timing, which is useful for diagnostic purposes as well as setting base timing.

the distributor that will work with many standalone EFI systems, although there are also versions with mechanical timing controls or with the same crank trigger used on Motronic systems.

Motronic EFI—The Motronic system replaced L-Jetronic and brought timing control into the ECU. Motronic systems have used everything from vane airflow meters to hot-wire mass airflow sensors to speed density with a MAP sensor inside the ECU.

Most Bosch Motronic systems use a 60–2 crank trigger with a VR sensor, often paired with a one-tooth sensor on the camshaft. Many aftermarket ECUs are set up to work with this sensor design. In fact, Electromotive designed their TEC family of ECUs to use this trigger setup exclusively. Bosch used the 60–2 wheel for distributor-based, wasted-spark, and coil-on-plug systems. Bosch frequently puts the ignition module inside the ECU, but they've also offered many external ignition module designs as well.

A few early Bosch Motronic systems used a flywheel trigger, where a VR sensor counted each tooth on the flywheel ring gear and a second sensor picked up a single reset tooth. This can work like a basic dual-wheel decoder, but not all aftermarket ECU designs are up to handling a primary input with over 130 teeth. Often you can update this design to the later 60–2 crank trigger by swapping on a crank pulley from a later version of the engine if you want to upgrade your engine management.

For idle air control, Bosch originally rolled out a thermal valve that needed no input from the ECU. Later designs used pulse-width modulated IAC valves, both single-coil and dual-coil designs. Bosch introduced one of the earlier throttle-by-wire

systems, which eventually supplanted idle air control valves on most of their current designs.

There are a number of companies that make performance chips for various Bosch systems, but devices that allow editing the factory ECU's settings in real time are harder to find. Because the sensors Bosch used are very similar across a broad range of makes and models, many aftermarket standalone ECUs can work with most of the original sensors if you want to replace your Motronic controller with something more tuneable.

Ford/Visteon EEC Systems

While Ford has often used Bosch injectors, they use their own computer and ignition designs. Their control computers go by the name electronic engine control (EEC). The EEC-I and EEC-II designs controlled emissions on carbureted cars. The EEC-III worked with feedback carburetors and a few throttle body injection systems.

Ford's EDIS distributorless ignition control module.

EDIS lends itself to transplanting onto vehicles besides Fords, as it can work with any engine that you can find a way to put a 36–1 crank wheel on. Here is a 36–1 wheel on a Mopar slant six.

Ford's EEC-IV, which ran from 1983 to 1995 in the United States and lasted even longer elsewhere, was their first system that really caught on with the enthusiast community. In 1996, the On Board Diagnostics (OBD-II) requirements prompted Ford to roll out a new EEC incarnation, the EEC-V, and in their newest models, as of this writing, you can begin to find the EEC-VI. Sometimes you will find a few products wearing the Blue Oval that used Nissan or Mazda engines and engine management, too. Today, Ford has spun off much of their electronics division into its own company, Visteon.

EEC-IV—The EEC-IV was the brain behind the 5.0 Mustangs, 2.3 Turbo, and the early modular motors, so many cars with the EEC-IV controller make prime hot-rodding material. All but the earliest EEC-IV computers used identical 60-pin connectors with similar pinouts, although the exact use of the pins changed depending on how many inputs and outputs the engine needed. Ford paired the EEC-IV with three different types of ignition systems. Ford designed all three of these ignition systems to work independently of the EEC-IV, so if they lost communication with the ECU, the spark still fired as normal except that the ECU was no longer in control of the timing. They put a

diagnostic connector in the wiring harness, and if you pulled the connector, the ignition would ignore the timing from the ECU and fire on a fixed timing.

Thin Film Ignition—Ford calls the EEC-IV distributor a TFI system, for Thin Film Ignition. These use a basic distributor trigger with a Hall effect sensor that sends the ECU a 12-volt square wave called the PIP signal. The ECU sends the TFI ignition module a 12-volt signal (called SPOUT, for SPark OUTput) back to tell it when to fire the coil. Some TFI modules controlled the dwell on their own, while others used the length of the SPOUT signal to control dwell. Sequential fire EEC-IV systems with TFI ignitions, like the later 5.0 Mustangs, almost always used one shorter trigger on the distributor for cylinder identification. There are a few obscure applications where they added a second sensor reading a single toothed wheel instead.

EDIS System—Ford later gave the EEC-IV a wasted spark ignition capability using an add-on module called electronic distributorless ignition system (EDIS). Rather than having the EEC-IV computer send separate signals to each coil, the EDIS module allowed the EEC-IV computer to keep to one channel of rpm input and one channel of ignition output. The EDIS module reads a 36–1 crank wheel and uses this to determine the crankshaft position. It then sends the EEC-IV computer a PIP signal that looks almost exactly like the signal from a TFI distributor. The EEC-IV sends it back a single signal called spark angle word (SAW). While normal ignition systems use the timing of a pulse to determine when to fire the coil, the EDIS system encoded the timing information in the pulse length. A shorter pulse length results in more spark advance. Controlling an EDIS unit requires special code in the ECU, but many aftermarket standalone systems have this.

The EDIS ignition system has taken on a life of its own in the low budget hot-rodding community. Because EDIS uses fairly simple wiring, and delivers a hot, accurate spark, many gearheads have transplanted it off of Fords and onto engines from other companies. You just need a controller, a way to fit the 36–1 wheel onto your crankshaft pulley, and a sturdy bracket for the sensor.

Motorcraft DIS—A few Fords used an intermediate step between EDIS and TFI. These have a distributorless ignition but lack the 36–1 trigger wheel. The module is marked "Motorcraft DIS" rather than "EDIS," and has two connectors where an EDIS module has one large connector. These appeared on the dual-plug 2.3-liter engine, the Thunderbird Super Coupe, and the Taurus SHO. This may look like an EDIS system, but its module uses similar signals to a TFI module.

EEC-IV computers come in both speed density and mass airflow flavors. The speed density systems use a somewhat unique MAP sensor; its output is a frequency instead of a voltage. Few aftermarket standalone systems read the Ford MAP sensor. Except for a few early systems which used DC servomotors for idle control, almost all Ford EFI systems have used pulse width modulated, single coil IAC valves, which are fairly easy to control for most aftermarket EFI systems.

EEC-V—Ford's EEC-V computer did away with the external EDIS module, integrating this module's components into the ECU. However, these designs still keep the 36–1 trigger wheel (or a 40–1 in the case of V-10 truck motors) and VR sensor. It's just that the coils and sensor wire directly to the ECU, with no module in between. EEC-V computers added more complex on-board diagnostics, and some of these systems support continuously variable valve timing control, ECU control over the gauges, and other more complex outputs than the EEC-IV used.

The popularity of the 5.0 Mustang ensured that the aftermarket would try to crack Ford's ECU codes. There are a number of chip tuners out there, such as TunerPro and the Tweecer line of tuning products, for tuning many of the EEC-IV and EEC-V line and even a few Ford-specific piggybacks like the Anderson PMS. Flash tools for OBD-II Fords are available from Superchips, Sniper, Diablosport, and several others. Standalone engine management system builders haven't ignored Fords either, with many American-built ECUs giving you the option of working with the factory sensors. Many companies even offer model specific wiring harnesses or plug and play systems for V-8 powered Fords.

General Motors/Delphi

General Motors first bought their EFI systems from Bendix. These systems controlled fuel only, and used unusual features by today's standards, like positive temperature coefficient temperature sensors and an ECU that switched the power to the injectors instead of ground. You can rewire the injectors to work with a modern standalone if you're dealing with a Bendix system, but usually the sensors will have to go.

It wasn't long before GM developed a new system in-house, although like Ford, they've now spun off their fuel system building to a separate company, in this case Delphi. Many of the sensors GM used, in

Ford's modular V-8. Note the coil-on-plug ignition.

The TPI system used on the L98 engine was one of GM's earlier efforts into high-performance, fuel-injected engines. The long runners make for a lot of low rpm torque, but the system is a bit restrictive at high rpm. There's an HEI distributor at the back of the manifold.

particular their temperature and MAP sensors, have become the default sensors in the aftermarket EFI world. They typically use stepper motor IAC valves, and there are even kits to put these IAC valves onto other engines from other makes because of the number of aftermarket EFI systems that cater to GM designs. However, many European GM products used Bosch engine management.

HEI Distributors—GM's first electronic spark control grew out of their mechanically controlled HEI distributors. The mechanically controlled HEI distributors used a VR sensor and mechanical timing controls, with a simple four-pin module firing the spark when it received a pulse from the pickup coil. GM reworked their HEI distributor by removing the advance mechanism and adding more connections to the HEI module, resulting in modules with seven or eight pins. One of the pins sends a square wave signal to the ECU when the VR sensor sends a pulse to the HEI module, and one of the pins lets the ECU fire the coil. These work like a basic distributor system, but GM adds an extra pin for what they call the bypass signal. When there is no voltage on the bypass signal, the HEI module fires whenever it receives a pulse from the VR sensor, ignoring any commanded timing changes from the ECU, making for more accurate cranking timing. The ECU sends a five-volt signal to the bypass signal to tell the HEI module it is taking control of the timing.

The LSx motors have tremendous mod potential. This Silverado makes over 1,000 horsepower at the wheels.

GM Crank Triggers—GM then designed a series of external distributorless ignition modules similar to Ford's EDIS. The first versions work with almost exactly the same signals as the HEI, so as to require as few changes to the ECU as possible. Unlike Ford, GM designs several different trigger wheel arrangements for their DIS systems, including a 6+1 crank trigger and several others. GM usually classifies their trigger wheels with a number followed by "X," the number standing for how many pulses the sensor gives per crank revolution. The number may not give an indication of how the pulses are spaced. For example, many engines have a "7X" sensor that has six teeth spaced 120-degrees apart and one tooth 10 degrees away from one of the others, the previously mentioned 6+1 crank trigger. The DIS module became integrated with the ECU in 1996 (also like Ford) but GM kept the same sensor arrangements. GM's crank triggers usually need specialized ECU code if the ECU reads the sensors directly, but you can sometimes use one of GM's external DIS modules if your ECU does not support your crank wheel, and then have your ECU interface with their DIS control module to control the timing.

Optispark Distributors—The LT-1 and LT-4 engines are an exception to GM's designs. For these engines, GM brought in a distributor design from Mitsubishi that most closely resembles a Nissan trigger wheel pattern, with an optical trigger. This produces two signals, the "180X" signal coming from a ring of 360 tiny slits and the "4X" signal coming from a ring of eight slits of different length. These eight slits of different lengths all have one edge that space out evenly around the disc; the other edge is uneven due to the different lengths. If you set your aftermarket EFI system up to read the evenly spaced side, then this works like a standard distributor as far as many aftermarket EFI systems are concerned, again, as long as you trigger off the evenly spaced edge. They also added a four-tooth

crank wheel when OBD-II appeared, which makes for a great trigger setup if you're keeping the distributor.

LSX Crank Triggers—The LSX engines introduced another type of distributorless ignition, although it appears to be related to the design GM used on the Cadillac Northstar. The original designs use a 24-tooth crank wheel and a one-tooth cam wheel, which you can use with an ECU that can decode a generic dual wheel. However, the teeth on the 24-tooth wheel are equally spaced only along one edge, so you must make sure your ECU is triggering on the correct edge. The stock ECU uses the irregular tooth lengths for faster cylinder identification, something that would take purpose-built software on an aftermarket ECU. The coils have built-in ignition modules. The LS2 and later LSX family engines use a variant of the 60–2 crank wheel with unequal length teeth. In GM terminology, it's called a 58X wheel, since it has 58 real teeth.

GM's ECUs are some of the most commonly hacked systems out there, with a proliferation of chips from big name companies to lone enthusiasts cranking out EPROMs in a spare bedroom. Software like LS1-Edit lets you tune some of their OBD-II ECUs in real time, and those looking to tune an OBD-I ECU in real time can often install a device like the Moates Ostrich into the ECU to allow it to be tuned in real time, too. Other options for the OBD-II systems include handheld programmers from companies such as HP Tuners, Diablosport, and Superchips. The popularity of the General's V-8s also ensures that there is no shortage of aftermarket standalone systems that are compatible with the factory sensors.

Chrysler

One of Alexander Pope's famous quotes is "Be not the first by whom the new are tried, Nor yet the last to lay the old aside." Chrysler has been famous for ignoring this advice, leading to both spectacular successes and spectacular failures. They introduced the first electronically fuel-injected production cars in 1958, but reliability problems meant that under a hundred of these Bendix Electrojector systems made it onto the road. The wax paper dipped electronics proved unreliable, and Chrysler soon recalled these cars and converted back to carburetors. Today there is only one known running Electrojector system in existence, surviving thanks to an electrical engineer who updated many of its components with more durable modern equivalents.

Lean Burn System—Chrysler returned to computer controls in the 1970s with the Lean Burn

Toyota's T series engine was an early recipient of EFI. Today you can still find heavily tweaked examples.

A 5.7 Hemi with the plastic cover off. The drive-by-wire throttle body and coil-on plug ignition are clearly visible.

system. While many cars in the 1980s used fuel injection with mechanical timing controls, Chrysler took the opposite approach and paired a carburetor with fully electronic timing controls. These used a VR sensor (or sometimes two VR sensors, one of which was used for start-up timing control) with one tooth per cylinder. While the Lean Burn computers often suffered reliability problems from their ECU mounting location (attached to the air cleaner) and extensive vacuum hoses, the distributors from these systems work very well when retrofitting EFI to older Chrysler engines.

Chrysler had better success with their injected K-cars, where they developed a new system of engine management suitable for mass-produced turbo motors. These used a Hall effect distributor. Throttle body injection versions used a basic distributor with four vanes. However, sequentially injected K-cars often had a "window" cut into the number-one cylinder's vane to provide cylinder identification, a feature not all aftermarket ECUs could work with. Like GM, Chrysler was often partial to stepper idle air controllers.

Reading Chrysler's trigger wheels often took purpose-built code. In the 1990s, Chrysler developed different trigger wheel patterns for the 2.2 SOHC, 2.2 DOHC, Jeep 4.0, Jeep 2.5, 5.2/5.9 Magnum, 3.9 Magnum V-6, the Viper V10, and 2.0 Neon motors. And that's not even counting several Mitsubishi based engines which themselves had unusual trigger wheels. While the 5.2 and 5.9 Magnums used a simple dual-wheel system, all of the others required an ECU designer who wrote code explicitly for the engine. Custom chips and factory ECU reflashes exist for many OBD-II Chrysler products from tuners like Superchips and Tom Fox.

Next Generation Controller—Around 2003, Chrysler finally standardized their crankshaft trigger wheel, settling on a wheel with 36 equally spaced teeth with two teeth missing and two teeth filled in combined with an irregular camshaft position reluctor wheel. They paired this with a new ECU design called the NGC, for Next Generation Controller. This unit employed a drive-by-wire throttle and replaced conventional fuel tables with a more complicated, model-based system.

Toyota

Much of Toyota's electronics come from their affiliate Nippondenso, which licensed technology from Bosch. Early injected Toyotas used a system called VAST for Variable Spark Timing, which resembles a GM HEI in that it combines a distributor trigger wheel with one tooth per cylinder with a VR sensor pickup and an external ignition module. Like the HEI, this ignition can keep firing if it loses its signal from the ECU. Also like HEI, the external module conditioned the VR pickup's signal to turn it into a square wave. Some VAST systems added an extra pickup in the distributor with a one-tooth wheel to enable semi-sequential injection.

Electronic Spark Advance—Toyota then rolled out a system called electronic spark advance (ESA). This one kept the distributor, but replaced the four- or six-toothed wheel with one with 24 teeth and made the second single-tooth trigger standard. Many of these systems use two VR sensors instead of one on the one-tooth wheel, to let the system sync up faster while cranking. It's often possible to feed the signals from an ESA distributor to a generic dual wheel decoder.

DLI Ignitions—Toyota built several variations on the ESA system, including a distributorless ignition called DLI and versions with 12-tooth crank triggers instead of 24-tooth cam triggers. Some of

Taking the cover off a Toyota DLI cam angle sensor reveals a one-tooth wheel with two VR sensors called the G sensors, and a 24-tooth wheel with a single sensor called the NE sensor. This setup shows up on most high-performance Toyota engines from the late '80s and much of the '90s.

Honda used a similar distributor setup throughout the '90s. This 2001 Integra was one of the last uses of this system.

the DLI ignitions used an unusual feature called *multiplexing*. These systems employed an external ignition module that received a timing signal from the ECU on one wire and a separate signal on one or two wires, telling the module which coil to fire. Another variant used on non-sequential engines used a 24-tooth distributor wheel and a four-tooth wheel, but the one-tooth wheel was absent. Toyotas in the mid-1990s began switching to a 36–2 crank trigger, similar to Ford's design but with a longer gap. These systems changed again in the mid-2000s to more complex wheels requiring specific decoding firmware. Their choice of idle air control has been all over the map, with thermal, stepper, single-coil PWM, and dual-coil PWM valves all putting in appearances at various times.

Variable Valve Timing (VVT)—Toyota adopted variable valve timing early on. Their early VVT systems used a basic on/off actuator to advance or retard the cam. The later VVTi system used a pulse width modulated solenoid to provide continuously variable valve timing.

It's very rare to see internal modifications on a Toyota ECU. Most tuners either use piggyback

systems or replace the factory ECU with a standalone engine management system.

Honda

Honda's first stab at EFI used the ECU to control just the fuel with mechanical timing controls. It wasn't long before they rolled out a computerized timing control system resembling the Toyota ESA. Honda's B, D, and H series engines all used a distributor with three trigger wheels, each with its own VR sensor. As previously mentioned, these distributors included a 24-tooth wheel that Honda documentation called CKP, a four-tooth wheel called TDC, and a one-tooth wheel called CMP. Aftermarket ECU installers can take their pick of running this as a basic distributor or as a dual-wheel setup. Some Honda models relocated the TDC or CKP wheel to a crank trigger and cut the number of teeth in half.

Honda's later distributorless ignition systems often used a 12-tooth crank wheel and a camshaft position sensor wheel with three irregularly spaced teeth. If your ECU does not have purpose-built code for this, it's possible to modify it to get a generic dual-wheel setup by grinding two teeth off the camshaft reluctor wheel, or go for an m-n wheel by modifying the crank trigger

VTEC—The name Honda is practically synonymous with the variable valve timing and lift electronic control (VTEC) system. The early VTEC designs used an on/off solenoid to switch between two different cam lobes rather than adjusting the cam phasing. The later i-VTEC system added the ability to continuously adjust the timing, too.

Nearly all injected Hondas used pulse-width modulated idle air control valve, and Honda also tends to favor speed density over mass airflow sensors. Like GM, Honda has attracted many

VTEC cylinder head on this Civic has an oil control solenoid valve that switches from a mild cruising cam profile to more aggressive lobes at high rpm. This gives the motor a broader torque curve than one with a single cam profile.

The S2000 ditched Honda's previous ignition setup for coil-on plug ignition.

Greg Amy's Nissan NX2000 race car being tuned on a dyno before competing at Road Atlanta.

computer geeks eager to hack the stock ECUs, but most of their effort went into OBD-I systems. However, OBD-II engines that use a distributor can often run on an OBD-I ECU with some rewiring or an adapter harness. Hondata, Uberdata, Neptune, and CROME are four of the major players when it comes to tuning software, with Hondata and Moates being two companies that build hardware allowing for real time tuning. Hondata also offers a programmer for some OBD-II applications.

Nissan

Nissan's first fuel-injected cars used Bosch L-Jetronic systems, but they introduced a different spark timing system on the 1982 280ZX Turbo that took on a life of its own. This system used a distributor with an optical pickup for computer-controlled timing. One ring of holes had six equally spaced slots, while a second ring of holes had 360. Nissan apparently used a dedicated piece of circuitry to count the 360 slots. With the introduction of the 300ZX, Nissan modified this system to use different slot lengths to provide cylinder identification, but the trailing edges were always equally spaced and can still be used without the need for a special mode designed for this wheel.

Nissan's system spread from the Z cars to other performance cars' engines such as the KA24DE, CA18DET, SR20DET, and the RB family of engines. This design spread to other Japanese and even American manufacturers, as this system formed the core of the GM Optispark. Four-cylinder motors replaced the ring of six inner holes with a ring of four holes. It's a rare aftermarket ECU that can read the 360 slots, so most ECUs simply use the ring of four or six slots. This is pretty

easy to use as a basic distributor trigger, but using this with a distributorless ignition requires code written specifically for the sensor, or modifications to the trigger wheel. Later Nissans used even more complex trigger designs, but most of them still have a fairly high tooth count. There are some options for tuning stock Nissan ECUs, such as Calum and Nistune for earlier systems and Cobb Tuning's system for the 350Z and GTR, but it's not uncommon to go to a Nissan race and find many of the cars sporting a standalone system, with or without aftermarket reluctor rings grafted onto the crank pulley to trigger such systems if needed.

Mitsubishi

As a large industrial conglomerate, Mitsubishi builds much of their own electronics, as well as electronics for other manufacturers. In fact, many of the optical triggers found on Nissans have the Mitsubishi tri-diamond stamped on them, although Nissan also bought some of these from Hitachi. Mitsubishi has been partial to optical and Hall-effect types of sensors and stepper idle air control valves for their own cars.

The earliest fuel-injected Mitsubishis, including the 4G54 powered Starions, used the computer just for fuel control and occasionally for knock retard. These early systems had a distributor with a mechanical advance and a VR sensor. Other Mitsubishis used the same optical trigger discs as Nissans, with one ring of 360 slots and another ring with one slot per cylinder.

4G63 Trigger—Mitsubishi came up with a different approach at about the same time they turbocharged their 4G63 and dropped it into the Eclipse. These use an optical sensor like the Nissans at first, and they're so close it is sometimes possible to swap trigger discs between this system and Nissan systems. The 4G63 trigger has one ring of four slots and an inner ring of two slots of different length.

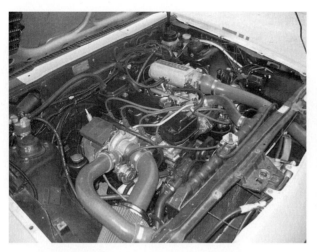

A Mitsubishi Starion with a G54BT motor, one of Mitsubishi's earliest high-powered turbocharged inline fours. Although this would have had throttle body injection from the factory, the owner has swapped over a multiport system from a later engine.

Some generic dual-wheel decoders can use this to control wasted spark, but sequential injection or coil on plug with this sensor requires special code to identify the cylinders. Later versions of the 4G63 used Hall effect sensors that created essentially the same pattern. DSMLink is one of the biggest names in hacking the '90–'99 ECUs for the 4G63, with other companies like EcuTeK offering flash solutions for later models. The latest Mitsubishis use more complex wheel patterns that would require firmware written specifically for them.

Subaru

To say that Subaru marched to a different drum would imply they still at least bothered to listen to someone else's music. Maybe at one point they did, but not so much any more. Subaru's designs now follow their own ideas of how to make the perfect car, and their electronics are no exception. Subaru's earliest fuel injection systems in the 1980s often used Nissan-like trigger discs, but when they went to distributorless ignition, Subaru came up with a unique reluctor wheel design for their flat fours having six teeth on the crankshaft and seven on the camshaft, both wheels having irregularly spaced

A late '80s Subaru RX uses a similar engine management system to early Nissans, with an optical sensor in the distributor.

Later Subaru motors like the one in this WRX used a distributorless ignition.

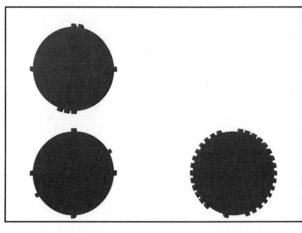

Two of the more common Subaru trigger patterns, the 6/7 at left, and the 36–2–2–2 at right.

teeth. Their flat sixes used a more conventional dual wheel with 12 equally spaced teeth on the crank and a one-tooth reset signal. More recent designs have combined an irregular two-tooth cam wheel with a crank wheel with 36 base teeth with several irregularly spaced missing teeth, known as the 36–2–2–2 wheel. Specially coded firmware is required to decode these irregular wheels, hence aftermarket ECU support for these wheels is more rare though support is growing. There is always the option of fitting a crank-mounted trigger wheel that's compatible with your ECU of choice. There are quite a few flash tuning options for the '99-and-later Subarus from Cobb Tuning, EcuTeK, and some others.

Mazda

Many of the Mazda sensor systems appeared on other Japanese and sometimes even American cars, but they did have a few unique systems. Their first fuel-injected cars used mechanical timing controls, adding an electronic knock retard on turbo engines. Others used simple optical or Hall effect distributors with one tooth per cylinder with a one-tooth cam sync. The distributorless versions of the 13B rotary engine use a Nippondenso trigger system similar to the design used on Toyotas, but with a 12-tooth wheel and a one-tooth wheel spinning at the same speed as the eccentric shaft.

Rotary engines also have two spark plugs per rotor that fire at separate times, requiring special ECU code to fully support this sort of ignition. The '90–'97 Miata used the same cam angle sensor pattern as the 4G63 in the turbo Eclipse, but later Miatas used their own design. Thanks to their partnership with Ford, there were also Mazda engines fitted with 36–1 crank triggers or even Ford TFI distributors, even when Mazda was not using a complete Ford engine outright, which also happened.

Some recent Mazdas have used the Subaru style 36–2–2–2 crank trigger. They have kept a fairly consistent preference for pulse width modulated idle air control system. However. Like Toyotas, there are very few options for hacking stock Mazda ECUs, although Cobb Tuning has a few options for the Mazda 3, Mazda 6, and RX-8 and a few specialty vendors offer chips for the MX-6. A standalone is more commonly the only course of action if you need more than the most basic levels of control that a piggyback may offer.

Mazda's rotary engines present their own unusual tuning demands. Each rotor fires once per eccentric shaft revolution, and there are two spark plugs per rotor, with the trailing spark plug firing a few degrees after the leading spark plug.

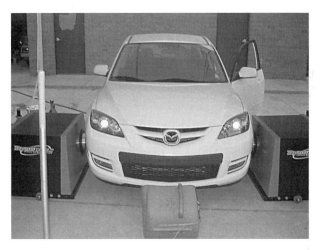

The Mazdaspeed 3 has a few off-the-shelf hacks available to reflash the stock engine management.

There aren't many options for hacking stock Mazda ECUs on earlier cars. Tuned Miatas often run MegaSquirt, Hydra, or AEM aftermarket ECUs.

Chapter 8
Tweaking the Stock ECU

A chip programmer can rewrite data in a memory chip. This particular one is for chips with flash memory, while other designs can write to EPROM chips.

Sometimes you don't need to change the whole ECU or fit a piggyback black box to retune a factory fuel-injected car. The original manufacturer designed their ECU so that their own engineers could tune it, and the aftermarket has sometimes found ways to tap into this, taking advantage of this to put their own calibrations into the ECU the car originally came with. While standalone aftermarket ECUs tend to be one-size-fits-all solutions, hacks to a stock ECU are by nature very model specific. Some ECUs have so many ways to hack them that anybody can buy a couple hundred dollars' worth of equipment and start cranking out "performance" chips in their basement, while other factory ECUs have encryption schemes so complicated that they have resisted all efforts to hack them thus far.

Hacking Methods

Hacks to stock ECUs take several different forms. The first such hacks came out in the 1980s, when digital fuel injection systems first became widespread and ECUs stored their data in an erasable programmable read only memory (EPROM) chip. With these early systems, you could sometimes open up the ECU, pull a single chip off the board, and then put the chip into a device that erased its memory and uploaded a new set of data. Some ECUs had the chip soldered to the board instead of in a slot, allowing it

to just be popped out, but it's often possible to change that with the appropriate soldering tools. With some ECUs, such as the OBD-I Honda, you need to add other circuitry to the board to allow you to put your chip in place.

In 1996, the US government required manufacturers to roll out a system called On-Board Diagnostics II (OBD-II). The main goal of this system was to detect if a car had malfunctions that made it pollute more, but one of its provisions had a side effect for the tuner market. OBD-II usually did away with storing engine tune data in EPROM chips and stored it in a type of memory that allowed reprogramming through a diagnostic port in most such ECUs. This served a pollution-fighting purpose in that if a manufacturer needed to recall a set of cars with emissions control calibration that wasn't working as expected, they could simply have their dealers upload a software update to the cars instead of having to pay the labor expense of removing tens of thousands of ECUs and swapping out their EPROM chips. It didn't take long for tuners to realize that what could be used to retune a car for emissions could also retune a car for performance. Again, just because it's OBD-II, doesn't automatically mean you can reflash it. These systems have to be reverse engineered (or one of the factory's engineers would have had to be persuaded to share a few tips on how it worked) to develop a product or software that could be used to reflash the system. But if that is available for your vehicle then you've got a fairly easy path to attaining at least some control of your engine without the trouble of a full ECU swap.

Nowadays, hacks to a stock ECU can take many different forms, from prepackaged tunes to devices that offer close to the tuneability of an aftermarket standalone ECU. Exactly what you can find will depend on what sort of vehicle you have. Because of the wide variety of tuning products out there and the way these products are often highly model specific, this chapter is intended more as a jumping-off point for you to further research for your vehicle, and how to tell if the options out there to hack your ECU actually meet your needs.

Devices that let you hack a stock ECU can be like the agents from *The Matrix*. They can bend many of the rules, and can have fairly significant power, but they are still bound by certain rules of the computer they are placed inside. Some tuning devices can give a stock ECU a few new inputs or outputs. Giving an ECU that only controls the fuel the ability to control the ignition timing, or turning a batch-fire ECU into a sequential fire system with only a chip change, just isn't going to happen.

"Boxed" Tunes

Prepackaged tunes or "tunes-in-a-box" come in both chip and flash devices, but they accomplish the same result: They replace the fuel and ignition maps inside the factory ECU with new maps created by a tuner. Most replacement chips come with a single tune, while a reprogramming device may allow you to upload several different possible maps. There are also devices that let you put more than one chip inside an ECU and change between them with an external switch.

The effectiveness of a prepackaged tune will depend on how similar the engine and related systems (cam, intake, exhaust, everything) and the tuner used to develop the map are to your setup. A ready-made chip can really shine when sold as part of a comprehensive turbo kit or along with a crate motor, so that the engines that use the chips are close matches to the engine the tuner used. The farther your buildup is from the beaten path, the less likely it is you will be able to buy a tune off the shelf that fits. Tunes for a TPI Camaro with popular bolt-on upgrades and a common cam grind are easy to buy. But put a set of Aardema overhead cam heads on that small-block and you'll have a very difficult time finding a tuner that has worked enough with such a rare beast to be able to provide a canned tune.

One other limitation on prepackaged tunes is that

OBD-II cars all have a standard port for communicating with the ECU. Originally intended for diagnosis, repairs, and recalls, there are many aftermarket devices that use it for data logging or reprogramming.

The Diablosport Trinity plugs into the OBD-II port and allows both tuning and data logging. Photo courtesy Diablosport.

since every engine is slightly different, tuners often do not try to push a tune as aggressively as possible, for liability reasons. Naturally, just how conservative the tune is will vary from tuner to tuner, but it's rare that you will see a ready-made tune designed to extract every last ounce of horsepower. On stock engines, you'll often see the largest gains on turbocharged engines where the ECU controls the boost. Engine to engine variations can be so big that it's not uncommon to see owners add a piggyback on top of a chip tune to get things dialed in.

If you're buying an off-the-shelf tune, take the time to research the tuner's reputation. Chip tuning companies range from internationally known brand names to guys burning chips on their kitchen table. A bit of research with other people who have the same car you have can help you find out the tuners who don't really have access to a dyno or who are incompetent enough to come up with chips that erase the code that lets the ECU control the air conditioner (or worse). It's rare for major name-brand chips to cause that kind of trouble, but even a chip tuner with full page ads in national magazines may not be all that good at actually delivering horsepower gains. Get references from a tuner's customers, ideally some who have put their

ECU hacks on parade. Most of these circuit boards are for installing in various stock ECUs to allow the user to reprogram the fuel and spark tables, or add features such as data logging. Many of these are intended for specific computers instead of one size fits all.

Honda's designers did not plan for anyone tuning their ECUs in real time, so giving an ECU this feature requires some extra circuits. The cables coming out of this ECU connect it to a ROM emulator and a set of PointStep gauges.

holding the engine in a single load cell. You can log data and correct your fuel tables based on this after running the engine in different sections of the table, but the ignition table you won't be able to dial in this way. You really need to be able to hold it in each load cell on a dyno and tune it in real time, but I'm getting ahead of myself. More on this in Chapter 12, the tuning chapter.

While making a chip can seem less expensive than buying a standalone system, getting the perfect chip requires considerably more dyno time than dialing in a standalone if you need to tweak the low load cells. The lack of real time tuning makes getting one perfect a time consuming process, and when it comes to dyno tuning, time literally is money, generally more than a hundred dollars an hour. If you aren't planning to sell copies of your tune, you may be better off spending a bit more to get an ECU you can tune in real time.

Tuning the Stock ECU in Real Time

For the tuner who wants to (or who the rules require to) use a stock ECU, but wants to be able to fully tune in all load cells at part throttle as well as wide-open, you may be able to find a real-time tuning option. As with chips, the availability is going to depend on the car. Some OBD-II ECUs can be tuned in real time, others cannot. You will need to check with your tuning device vendor to see which ECUs it can tune in real time. Chances are you bought it with a specific job in mind.

While an out-of-the-box OBD-I ECU will not lend itself to real time tuning, you can get around this by putting more devices into the box. Replacing the ROM chip with an emulator like the Moates Ostrich will give you a tuneable chip that can be adjusted when the engine is running with immediate response to your changes to the tune. There are sometimes even emulators for OBD-II ECUs when such ECUs cannot be tuned in real time. For some ECUs popular with serious hardware hackers, you can order a kit that you solder into the ECU at home. In other cases, you may need to ship your ECU out to a tuning shop and have them rework it to allow tuneability. Depending on your car, you may actually need to swap to the stock ECU from a slightly different model, such as replacing a Prelude ECU with one from a Civic.

Limitations on Hacking the Stock ECU

Sometimes, it can be difficult to decide if you want to hack a stock ECU or convert to a standalone system. It may seem like hacking the stock ECU could get you all the things you'd need

chips to good before-and-after tests on a dyno or at least a dragstrip, before ordering a mail-order tune.

Home-Brewed Chips

These days, making performance chips isn't just for companies like JET or Superchips. Thanks to the cooperative efforts of several tuning communities, it's now possible to make chips for '95-and-earlier GM and Honda products in your own garage. Making the chip itself generally requires two things—an EPROM burner and the editing software for a PC.

Homemade chip tunes can be tricky to optimize, however, since typically you can't tune them in real time (though in some cases you can with additional emulator hardware). Getting a chip tune dialed in right is an iterative process, making a map, testing it with a dyno pull, then examining the data readings to create a new map and finally, trying the new one out. While you can often fairly quickly get the air/fuel ratio adjusted mathematically, the spark map is going to create a stack of chips to do properly, and without specialized hardware to allow real-time tuning, you'll struggle to get the part throttle ignition fully optimized. At full throttle, the tuning process is very much like full throttle pulls with a standalone; getting the fueling on target with a couple pulls and then adjusting the spark to see what effect your changes have on power output, creating a stack of chips along the way. Dialing in part-throttle drivability can be tougher without the ability to adjust fuel and ignition, while out of a standalone, with less money. Well, sometimes you can, and sometimes you can't. Here are some limitations that you may run against when using ECU tuning devices.

The first limitation is that for some cars, a hack to tune the stock ECU just plain isn't out there. While there are more tuning options for GM LS1s and Honda B series motors than you can shake a timing light at, finding one for a limited production vehicle, or even for many other commonly modified sports cars just isn't always an option. Sometimes the aftermarket is the only way to go.

The second is that you are mostly limited to the features the stock ECU had. Sometimes you can take over circuits that originally served functions a race car does not need, or even populate circuits that the original designer came up with but left incomplete, and use these for purposes the factory never intended. For example, Hondata and some other tuning tools allow you to put a boost control circuit into an OBD-I Honda ECU, even though Honda never put that particular ECU on a turbo car. You'll also sometimes see nitrous control or table switching inputs. However, sometimes there are features you just plain can't add to a stock ECU. For example, if you have a Bosch L-Jetronic ECU that only controls the fuel and has little influence over the timing, you're not going to be able to add a spark control map to it, or turn it from a batch-fire ECU into a sequential one. Removing a restrictive airflow meter can also be difficult when you're hacking the ECU, although some stock ECUs allow changing its fuel control method. Some ECUs have more subtle limits, such as not allowing you to enter an injector size beyond a certain maximum number. This can limit that maximum power you can support with the stock computer, even retuned.

The third is that factory ECUs were originally designed for factory development programs with multi-million-dollar tuning budgets. Consequently, some tuning programs may have rather steep learning curves, or lock you out of many of the areas of the program. Being locked out of esoteric settings, like how the ECU models inlet port temperature may just make your job easier if you're trying to tune your engine for some mild bolt-ons and a cam swap. This could be something of a nuisance if you've replaced your iron heads and intake manifold with aluminum heads and a carbon fiber manifold.

Speaking of nuisances, and the lack of multi-

If you have a Honda "P28" ECU (identified by part of the code on the label), there are quite a few different ways to hack it. This ECU can be swapped in places of many OBD-I Honda ECUs, and even OBD-II ones with an adapter harness. Not all the OBD-I Honda ECUs are hackable, and hacks for OBD-II Hondas are much rarer.

Inside a stock Honda P28 ECU, there are several pads for mounting chips the factory did not install in this application. Most hacks to this ECU involve populating these with either a user-tuned chip or a ROM emulator.

million dollar budgets, many efforts to hack stock ECUs start as efforts by a single enthusiast who also happens to be a software programmer. While some of them grow from these roots into bigger commercial developments, other projects remain a one-man show. This can spur some of the best motivation for innovation and development. There are times that the end result is an application that maybe only the original programmer ever fully understands. In other words, it may have a very steep learning curve, not always do what you tell it to do, and frustrate you to no end as you desperately try to tune your hacked ECU on the dyno and can't get the changes you've made to take effect. The steeper learning curve mostly comes from figuring out how to work around these quirks. Thankfully this is the exception, not the rule. And you're guaranteed to learn something through the process. Even if it's just that you want to use a standalone instead.

As for what options may be available to you and what successes have been had, the best test may be to check with your enthusiast community. See if serious tuners who are running power levels like the ones you hope to reach and have the resources to dyno-tune a car properly are using tweaked stock ECUs or deep-sixed the factory parts in favor of aftermarket units as soon as the car arrived in their shop. Also find out how satisfied they are with the tuning product's ease of use and its supplier's customer service. Checks like these can help you find where the stock ECU reaches its limitations so you can determine the right course of action for your ride: hacked ECU, piggyback ECU, or full standalone.

Chapter 9
Piggybacks and Fooling the Stock ECU

Adjusting a MAF sensor's output curve will change the fuel and timing. This ABACO MAF is more tuneable than most, as you can adjust its output curve.

Piggyback devices change the signals going to or from the stock ECU without altering the ECU's inner workings at all. These little black boxes of electronic trickery often give a limited range of adjustment, and sometimes are referred to as the Band-Aid solution of the tuning world, because they do not offer as much control as a standalone or a device that lets you edit the stock ECU's tables. But just as minor injuries don't require a scalpel and stiches, sometimes a lightly modified car may not need a lot of adjustment from the stock settings. Piggybacks are often cheaper than standalone systems. Plus, they're often pretty straightforward to install, and if the car already runs on the stock ECU, you won't have to spend time trying to figure out how to start the car on a standalone ECU and can go straight on to tuning with the benefit of the stock ECU's programming and just use the piggyback to tweak it a bit.

Not all piggybacks can make the same adjustments, however. Some piggybacks are more limited and only control fuel, or only control timing, and sometimes they can only make adjustments in one direction (only add fuel, or only retard timing). A few are even more specialized, such as devices that prevent a speed limiter from kicking in. Sometimes it may take several separate piggyback boxes to give a car what it needs to perform well. Other times you can buy a single piggyback device that includes all these functions in one box.

Non-Programmable Piggyback Devices

We'll first cover piggyback devices that can (sometimes) be adjusted, but aren't truly programmable. These are more like the main jet on an old carburetor, having one single adjustment that applies across the board instead of a map that sets a complex set of different conditions.

MAF Tunes—One common way to get an ECU to run larger injectors is to swap out the mass airflow sensor with a MAF sensor using a different calibration curve. If your injectors flow twice as much fuel, the replacement mass airflow sensor (or the black box intercepting and modifying the signal from the MAF) tells the ECU the engine is flowing half as much air. The ECU will respond by cutting the pulse width in half, and the air/fuel ratio returns to where it should be, at least in theory. As the ECU often is programmed for a different air/fuel ratio at lower throttle amounts, so the fuel curve may not be quite ideal unless the tuner has carefully taken this into account. This also doesn't account for settings that are not calculated based on the MAF sensor, such as cranking pulse widths or tip-in enrichment. These are both examples where going to bigger injectors will still flow more fuel even, with a MAF signal adjusting piggyback in place, and they can affect drivability.

Besides the limit on the fuel curve and other potential fueling issues, swapping the mass airflow sensor has another disadvantage: It skews the timing curve if the ECU controls

The Infamous "eBay Chip"

One device that falls into the category of piggyback device isn't merely limited; it is so ineffective that reputable tuning companies will almost never touch it. This device has picked up the nickname "eBay chip" because many of its promoters sell it on eBay rather than establish their own sales channels. The sellers sometimes call it a "chip" even though it is not actually a microchip. Recently the sellers have shifted their terminology; the most common term nowadays seems to be "performance module."

The ads for this device often make promises even more outrageous and less believable than a politician's campaign promises. It's not uncommon to see the sellers claim this device can give 30 horsepower, and some go so far as to claim their device gives 70 more horsepower and 25 more miles per gallon. It's very rare for an otherwise stock car to pick up 30 horsepower with a full standalone system and a master tuner tweaking it unless other changes were made to bring that power out, and 70 hp is something you'd only see with computer mods alone, if the ECU controlled a turbo that had enough airflow capability to allow big gains in boost. The only secret to their tune is that there is no limit to how much a performance gain you can advertise when you are devoid of any semblance of a conscience.

In reality, this "chip" is nothing more than a resistor spliced inline with the wire leading to the coolant temperature sensor that makes the sensor read at a colder temperature. This causes the ECU to go into warm-up mode, making it add more fuel. This would only give a horsepower gain if the engine were tuned to run lean from the factory, but most motors are just the opposite, and the extra fuel would hurt power. However, many factory ECUs also add a bit of timing when the engine is cold, which may increase power as the factory tune is usually rather cautious.

Gain up to 50 hp!!!
+10 mpg!!!
-0 truth!

LYING BASTARD

LB

PERFORMANCE CHIP

PERFORMANCE MODULE

Worthless "performance modules" typically look something like this. There's seldom anything more than a resistor inside that box.

Since these two effects work at cross-purposes, these "chips" are as likely to hurt power as to increase it, and adding extra fuel definitely isn't going to help with gas mileage.

Of course, you'll also see people selling real performance chips on eBay, from products by big-name tuners to backyard EPROM burns and the tuning equivalent of pirated software. Just because it's a chip and on eBay does not automatically mean it's junk. Consider the tuner's reputation before buying a chip, whether it's on eBay or anyplace else. If you already own one of these, go ahead and put it in the trashcan alongside what's left of your eBay "electric supercharger," along with all of the fan blade bits that lodged into your intake after it grenaded.

the ignition. Using the MAF signal to cut the injector pulse width in half will also cause the stock ECU to think the engine is running at less throttle and advance the timing. Because of this, combining a MAF tune with significantly larger injectors means you will need a different means of controlling the timing, either a separate timing controller or hacks to the OEM ECU. Oversize mass airflow meters often pair better with chip tunes or devices that edit the ECU settings, instead of being a complete tuning solution on their own.

A few MAFs like the ones from ABACO have built-in memory that let you adjust their output function. This has the same limit on timing control, but gives you more flexibility than going with someone else's predetermined MAF calibration.

Clamp Devices—Factory ECUs often put certain limits on your driving. It's most common to see an

ECU set to limit the maximum rpm or maximum speed, but you'll also see some ECUs on turbo cars that cut back the engine's output if the boost gets too high. While single purpose piggyback devices have a much easier time adding rev limiters than removing them, there are single purpose devices for removing speed limiters or boost limiters. These are known as clamp devices because they clamp a signal when it reaches a certain value. That is, they hold the signal at that value and do not let it get any higher.

Speed limit removal devices intercept the signal from the vehicle speed sensor. When the car goes faster than the speed limiter will allow, the device sends back a false speed signal that's slower than the maximum allowable speed. Most of these devices are vehicle specific; some may also need to tamper with a signal from the transmission so the ECU thinks it is in a lower gear too.

An MSD Boost Timing Master is a piggyback timing control for distributors. The knob at right lets you adjust how much timing it pulls out as boost increases.

An add-on rev limiter module. Photo courtesy Pertronix.

Trying to remove an overboost fuel cut is somewhat riskier. A device called a *fuel cut defensor* intercepts the signal from either the MAP sensor or mass airflow sensor and prevents it from going over a value that would trigger the fuel cut. This has one obvious problem: It also prevents the ECU from accurately metering air and adding more fuel. For this reason, you should never use a fuel cut defensor on its own. These must be combined with some sort of device that lets you maintain the correct air/fuel ratio under boost. You'll often see this sort of clamp integrated into more sophisticated piggybacks that can control the injectors directly or operate additional injectors.

Simple Spark Piggybacks

Aftermarket ignition companies have come up with quite a variety of spark controllers that let you add functions by intercepting the ignition signal. These usually just wire up to the distributor, but they can work with almost any sort of engine management that uses a distributor, or even carbureted cars. Many of these are, in fact, designed for cars that did not originally have any computers, anywhere. But they can work on injected cars too.

One popular example of a timing adjuster is MSD's Boost Timing Master. This little gadget has its own MAP sensor and splices into the wiring to the coil. It responds only to positive manifold pressure, and retards timing under boost. A dial lets you adjust how many degrees of timing to pull per pound of boost. There are a couple of similar competing devices out there. Some versions, like the MSD 6-BTM, combine a boost retard with a capacitive discharge coil driver, letting you both adjust boost timing and get a hotter spark with one box. There are a few boost retard devices out there for distributorless ignitions, too.

A related set of devices accomplish the same thing for nitrous. These retard timing by a fixed amount when you activate a nitrous system. Both boost and nitrous retard boxes work only one way in most cases: They can only retard timing and not advance it.

Many aftermarket ignition boxes offer an adjustable rev limiter. These usually cut off sparks in a rolling pattern, firing some cylinders but not others so as to avoid kicking in too harshly. MSD's original rev limiter required swapping out "pills" (actually small resistors in a plastic module that looks sort of like a fuse) but newer designs let you adjust the rev limit with a knob. These can't usually raise the factory rev limiter, only let you add one of your own.

One of the newest tricks to show up in a piggyback ignition module is what's known as *slew control*. This retards spark timing or even applies a rev limiter if the engine accelerates too fast, in a move to limit wheel spin (since the engine can accelerate a lot faster if you're smoking the tires). It comes very close to working like traction control, but often gets around sanctioning body rules that ban traction control since these rules often ban systems that monitor wheel speed directly.

Peak-and-Hold Injector Drivers

If your stock ECU (or even an aftermarket engine management system) is built for high-impedance injectors and you want to run low-impedance injectors, you may not have to replace the ECU. There are several black boxes on the market that you can splice in between the ECU and the injectors that take the signal from the stock ECU and convert it into a peak-and-hold signal that drives a low-impedance injector. These don't change the engine's tune itself, so you will need a way of adjusting the ECU if you've changed the injector size. But if you already have a good means of tuning your fuel curve and your current ECU won't drive high impedance injectors, a peak-and-hold driver box can save you from the expense of a different ECU.

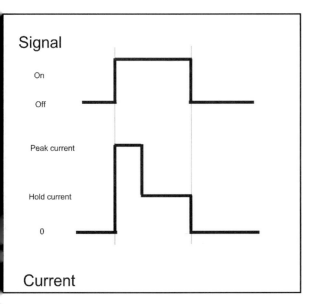

Peak-and-hold injector drivers change a continuously on signal into a two stage signal to allow a large amount of current to open the injector and a lower hold current level to keep the injector open.

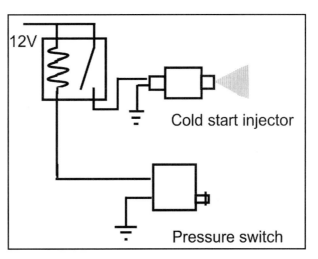

An on/off extra fuel device turns on a fixed amount of extra flow above a preset boost point. This can work, but it's not going to maintain a correct air/fuel ratio across the rev range.

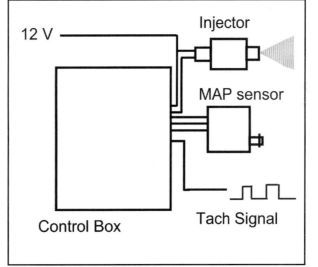

A more sophisticated extra fuel device is essentially a stripped-down version of a standalone ECU. It continuously adjusts the fuel delivery based on boost and rpm, allowing for better control of the air/fuel ratio.

Extra Injector Controllers

If your original fuel system can't supply enough fuel to meet a modified engine's needs, one piggyback approach is to simply add more injectors and a separate control system. You might need a larger fuel pump, but you can leave the stock ECU and injectors alone.

The oldest version of this method of adding fuel was used on carbureted turbo motors. It just used a boost activated switch that opened a valve to let gas flow through a small restrictor orifice and into a nozzle in the intake, not even a proper fuel injector. It flowed the same amount of fuel regardless of boost level or rpm. A few later systems used a pressure regulator that added more fuel with increased boost, but it still did not respond to rpm. While such a crude device can sometimes stave off detonation, it's only slightly more feasible than having a trained woodchuck pour gas down the throttle body from a pitcher, since it's really hard to train a woodchuck to stay under the hood with all the noise and heat. Otherwise, this is not much more effective.

At the other extreme, you could use a standalone engine management system just for control over the extra injectors. This can be expensive, since the system has enough computing power to run the engine by itself, but it can respond not only to rpm, but to changes in intake temperature, altitude, or anything imaginable.

A few devices fall into a middle ground, giving decent mixture control at a more reasonable price.

An extra injector controller can get away without handling such tasks as engine warm-up or acceleration enrichment, so it doesn't need to be as smart as a full standalone. A few companies offer stripped-down control units with simplified programming and minimal inputs, often just a MAP sensor and an rpm signal, although they may also have an air temperature sensor or a few other inputs. These typically use a fuel map similar to one from a full standalone system.

Obviously, these controllers do nothing about the ignition timing. And they don't let you lean an engine out if it is running too rich. They have a single job, to add more fuel. Usually you'll see these offered with turbocharger or supercharger kits as part of a complete package.

Programmable Input Modifiers

Now we come to devices that start to be tuned a bit more like a standalone. Devices in this class typically intercept the signal from the MAP sensor or mass airflow sensor before it reaches the ECU,

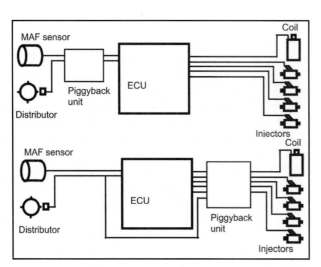

The wiring difference between modifying inputs (top) and outputs (bottom). Modifying the outputs can make it easier to tune the fuel and ignition separately instead of having one change modify both at once. Note that output modifying boxes may sometimes also tap into the input wires, but may not end up changing these signals.

and let you substitute a new signal. These have a programmable table of how to modify the sensor signal, similar to a standalone system's tuning tables. Apex'i built the most widely known input modifier, the Super AFC series, but there are other versions out there like the Greddy eManage Blue (although the eManage can be expanded to more than just an input modifier). The Jet Performance Module (not to be confused with their chips) is sort of a non-programmable version of this with a fixed tune.

Once you get to piggyback systems with this level of sophistication, you may find they have other functions built in. Some input modifiers can take a signal from a MAP sensor and create a fake mass airflow signal, allowing you to remove the airflow meter. This can help get around problems caused by a restrictive meter, or one that simply can't measure enough airflow for a heavily modified engine. It's also possible for these types of devices to have programmable on/off outputs.

A programmable input modifier is often easy to install, often requiring just a few wiring splices. However, it gives you a very limited range of tuning, and unless your ECU only controls fuel and not timing, attempts to change the fuel curve affect the timing curve. Setting one to reduce the MAF signal by 50% may not reduce the fuel input by 50%, but it also will cause the ECU to advance the timing by a rather unpredictable amount. It also has

An Apex'i Super AFC Neo. One of the most commonly seen piggybacks, the AFC line of products modify the MAP or mass airflow sensor readings.

trouble tuning such things as acceleration enrichment. Acceleration enrichment often uses the rate of change of a signal rather than the signal value itself, which can complicate tuning.

On some engines, tinkering with the input has another, sneakier problem. Newer factory ECUs are meant to compensate for wear and tear, and that includes worn sensors drifting from their factory specs. On some cars the ECU may interpret the modified input from a piggyback as a sign that the mass airflow sensor is drifting out of calibration, and find a way of returning to its old maps. This is not a problem for every engine, and it's more likely to be an issue with newer and smarter ECUs.

It is possible to adjust the timing by shifting the inputs to the ECU from the crank angle sensor. This can let you tune out any changes the MAF signal modifications make to the ignition timing and tune in what timing changes you need, if your piggyback device supports this. ChipTorque's XEDE is an example of a high-end input modifying piggyback that allows changing timing in this manner.

Programmable Output Modifiers

This category includes some of the most powerful piggyback controllers on the market, such as the Greddy eManage Ultimate and the AEM F/IC. A well-designed piggyback system that generates its own signals to the injectors and ignition has close to the power of a full standalone when it comes to tuning. These devices often have the ability to modify inputs too, but they usually only use these as a sort of built in clamp device to disable a speed limiter or overboost protection, or to give the ability to switch an engine from mass airflow metering to speed density. Some of them can even override a factory fuel cut rev limiter.

Not all output modifying piggybacks are created equal, however. A few units like the PowerCard can only lengthen injector pulse width, not shorten it. Since it cannot subtract fuel, such a system would not be able to work with larger injectors than the stock size without making the engine run too rich. There are versions of the PowerCard that instead allow you to control extra injectors in place of larger injectors in the stock location. This is still a "one way" sort of tuning, but can be a fairly straightforward way of adding more fuel than a stock fuel system can support. Other piggybacks, like the current AEM F/IC, can retard timing but not advance it.

A good output-modifying piggyback can take the ECU's pulses to the injectors and the ignition system and replace them with pulses generated by the piggyback. This lets you control the ignition

MSD's Timing Twister is a spark-only piggyback controller available for a variety of late-model American motors. Timing adjusts with a screwdriver instead of a laptop.

TIMING TWISTER
'96-'04, FORD
PN 86252

AEM makes a plug in harness to make it easier for its F/IC piggyback to tap into the stock harness and intercept the signals leaving the ECU.

and fuel separately, where an input-modifying piggyback would usually change both if it changed a sensor reading. Typically you can adjust the pulse width plus or minus 100% of the original value and have a measure of control over the timing too, although just how much control you have depends on the unit. A good piggyback of this sort will also let you make acceleration enrichment changes.

A serious piggyback from this category tunes much like a standalone ECU. You hook it up to a laptop and use tables for fuel and spark. The main difference is that the tables represent how much to alter the fuel or timing, instead of tables that generate these numbers directly.

This sort of piggyback has much of the power of a standalone ECU, but it still has to play by the factory ECU's rules for the most part. For example, they usually cannot get past a spark cut based rev limiter, and the methods they use to get around a fuel cut rev limiter do not give the same level of fuel control above the factory rev limit as a full standalone system. But they offer close to the same amount of tuning ability if one is available for your vehicle, often at a lower price tag. Keep in mind as well that you'll still be limited to your stock ECUs inputs/outputs, and any that the piggyback itself might add. So if you're looking to add boost or nitrous control to whatever else and want your EMS solution to control it, make sure you've got that covered. These are features that are commonly supported by standalones, not so much in the piggyback space, however.

The Math of Piggyback Acceleration Enrichment

Acceleration enrichment depends on the rate of change of a signal, rather than the signal value itself. If the ECU uses the same sensor for both load and acceleration enrichment calculation, such as some early Bosch Motronic systems, you're pretty much locked out of adjusting the acceleration enrichment. If you use different sensors for different tasks, such as a mass airflow meter for primary fuel calculation and a variable throttle position sensor for acceleration enrichment, it's still tough to alter the acceleration enrichment.

Consider the simple case of an ECU that had a throttle position sensor that goes from 0 to 5 volts, and adds an amount of acceleration enrichment that is directly proportional to the rate of change of the TPS reading. And suppose we want to give the engine twice as much acceleration enrichment. To get twice the rate of change, you would need to give twice the TPS signal, only this would mean you run out of range halfway through. If you opened the throttle halfway, the piggyback could adequately adjust the acceleration enrichment. The piggyback would be sending its 5v signal to the ECU, which is the maximum value, but you're only at half throttle. That's where the problem comes in, because now if you stab the throttle again and opened it fully, the ECU would already think the engine is at full throttle and wouldn't see the throttle stab at all, giving no acceleration enrichment. Cutting the acceleration enrichment in half, at least, would be a bit more workable.

In reality, a factory ECU will probably use a more complicated way of determining acceleration enrichment. While some just use throttle position, many factory ECUs also consider the engine rpm and sometimes the primary load sensor reading as well. If the factory ECU uses a sophisticated program that tries to factor in how fuel adheres to the intake walls, trying to change the acceleration enrichment becomes just about impossible. So most input modifying piggybacks just get tuned for steady state conditions.

You can see that piggybacks have their challenges and sometimes limitations that have to be worked around or lived with. These typically become a bigger issue the further you stray from the factory hardware the stock ECU was controlling. If you want an accurate way to adjust the acceleration enrichment, it's hard to do this without a device that controls the injectors in a more direct manner than altering sensor inputs.

Chapter 10
Standalone Engine Management Systems

The Thruster is Accel's current entry level standalone ECU.

Standalone EFI controllers can give you as much tuning flexibility as you could possibly need. These are designed to be installed on a wide variety of engines rather than built to run one particular motor, and as such they can tune your engine to work with practically any modification you're likely to throw at it. They are designed to make the tuning process straightforward, and it's rare to be locked out of any of the settings. Some standalone ECUs even make the source code available so that a determined programmer could reprogram, adding features he could dream up while on the race track. Some closed-source systems may have a support team that can add user-requested features for the right price. What you can do with a standalone is usually limited primarily by its available inputs and outputs, and since these were usually designed to go in a race car, you should be able to find a system with all the inputs and outputs you need to get your car ready for race day. While some cars/drivers have somewhat limited needs for extra inputs and outputs, others envision all sorts of control and data logging capabilities and will benefit from a controller with the capability to manage this.

If you're going for a standalone EFI controller, it's easy to find the number of options out there intimidating. The goal of this chapter is to help you narrow down your options. We've listed many of the most common engine management systems in the North American market, with a basic rundown of their features as well as some of their unique quirks and other information to help make an informed choice. These ECUs span quite a range of target markets, from some that target the professional racer to others meant for a backyard mechanic who'd like to get an old hot rod swapped from a carburetor to EFI with as little time and money as possible.

Accel DFI

While DFI was not the first company to offer a user-tuneable engine control unit, it is one of the earliest such companies that's still around. Since DFI was bought by Accel, they also get to be first on this list of ECUs, which is in alphabetical order by manufacturer. Their current systems are the Generations 7 and 8, and the Thruster, an entry-level system.

The current versions of the DFI offer sequential injection using speed density or alpha-N fuel metering. The Generation 7 and Thruster support distributor-based ignitions along with Ford EDIS, the GM external DIS module, and LS1 applications. The Generation 8 adds support for more distributorless ignitions, including the Chrysler Hemi. The list of extra outputs varies; the Thruster has outputs for one stage of nitrous, two-step rev limiting, torque converter and AC control, and fan control. The

A screenshot of Accel's tuning software. The entry level version allows access to basic tables, while unlocking more advanced tuning options requires a Pro Key dongle.

AEM markets both universal and plug and play standalone engine management systems, as well as piggybacks.

A BigStuff3 Gen 3 ECU. The one in this picture includes the optional transmission control output.

Generation 8 has extra fan outputs, support for progressive nitrous, VTEC, and boost control. Accel also offers turn-key EFI conversion packages, dual-sync distributors, and model-specific wiring harnesses. They also manufacture quite a few intake manifolds for Chevy and Ford V-8s, and a throttle body injection unit.

One of the quirks of the Accel system is the pro version. The normal version has some of the tables locked out, generally settings an inexperienced tuner is not likely to touch. Buying a dongle from Accel allows you to unlock access to these settings.

AEM

AEM offers a wide variety of universal, plug and play, and even piggyback fuel controllers. Their engine management system runs towards the higher end of the pricing scale, but with a fairly broad feature set as well. Their Series 2 system includes up to 12 injector drivers and 8 ignition drivers, and it supports EGT feedback, boost control, nitrous control, variable valve timing control, many general purpose inputs and outputs, and some of their plug and play applications even have drive by wire throttle control. Fuel control is sequential, with individual cylinder trim. It can use speed density, alpha-N, or mass airflow fueling.

One interesting option for the AEM system is their Engine Position Module. This is a drop in replacement for the distributor, and contains a trigger wheel with two optical sensors, one reading a 24 tooth wheel and the other reading a one tooth wheel. It's available for many Honda, Ford, and Chevrolet applications, and generally intended for

people who want to swap out a distributor for a coil-on-plug system. As this is quite a common wheel pattern, the engine position module could be used with many other engine eanagement systems besides AEM.

AEM was one of the first manufacturers in the plug-and-play market, and their catalog covers quite a range of applications. Their lineup includes many Honda, Toyotas, and Mitsubishis, as well as the 5.0 Mustang, Dodge Viper, Nissan 240SX, and some versions of the Subaru WRX. AEM also makes plug-and-play piggyback systems in addition to standalones.

BigStuff3

BigStuff3 was founded by Jon Meany, the same engineer who designed the original version of the Accel/DFI system. It built two ECU product lines, the Gen 3 for racers and tuners, and the Gen 4, a research system whose price and complexity means you're not likely to see a Gen 4 outside of an engine lab. We'll focus on the Gen 3 in this book, as if you are using a Gen 4, you probably already have a library stocked with college-level books on engine theory.

The Gen 3 is a sequential-fire ECU that can use speed density or alpha-N fuel metering with 16 x 16

The original Bowling & Grippo ECUs such as this MegaSquirt V3.0 were designed to be sold in kit form and installed in an off-the-shelf industrial electronics case.

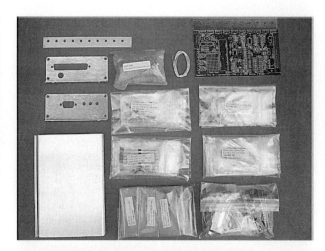

Bowling & Grippo distributors have built several plug-and-play devices based on the Mega-Squirt line. This is a homemade adapter for Ford EEC-IV units installed in a stock EEC-IV case, with the MegaSquirt on the right.

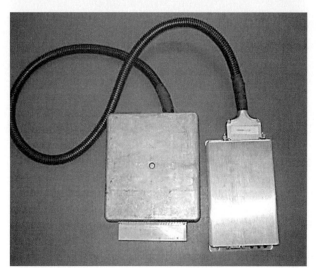

Bowling & Grippo/MegaSquirt

Bruce Bowling and Al Grippo developed the MegaSquirt and several related EFI products as a way for hobbyists to learn about engine management and build an inexpensive ECU at home. Their products follow a somewhat unusual distribution model, as Bowling & Grippo originally only sold a few key components of the ECU such as the circuit board and processor. The buyer would then source the remaining components and solder the ECU together using directions posted on the Internet. They also made the source code available so that computer savvy owners could alter it to fit their needs. It wasn't long before distributors sprung up who offered the MegaSquirt as either a solder it yourself kit including all parts that you might need helping you avoid the run around previously required, or for those not desiring to start an electronics project, you can purchase a turn-key ECU. Some newer Bowling & Grippo products are only sold fully assembled, but most of the assembled MegaSquirt products are actually built by the distributors using parts obtained from Bowling & Grippo.

MegaSquirt was not the first engine control system distributed this way, and several others have tried similar methods on subsequent ECUs. However, the Bowling & Grippo products are certainly the most widespread and successful ECU using this approach, having a very wide user base and excellent support resources.

Original MegaSquirt Features—The original MegaSquirt was a simple batch-fire fuel controller that operated little more than two banks of fuel injectors, a fuel pump relay, and an on/off idle air control valve. This didn't take long to change, as Bowling & Grippo encouraged people to modify the code and the hardware. Users soon added distributor-based ignition control, and eventually expanded the MegaSquirt's hardware to support as many as six ignition output channels and many different sorts of inputs and outputs. Speed Density, Alpha-N, Mass-Airflow, and a couple forms of the blending of two of these load measurements are supported.

One aspect of MegaSquirt that can make its learning curve a bit steep is that it doesn't quite have a standard feature list. The majority of MegaSquirt ECUs get put together by home builders who may add or omit features as they see fit. Even systems purchased from a dealer are often built to order and usually don't include every option in the book. The one constant feature of the original MegaSquirt design is its support of two banks of fuel injectors in a batch-fire configuration.

tables. It permits individual cylinder fuel and spark trim using a fixed percentage for fuel and a fixed trim angle for spark. One optional output the Gen 3 has that is seldom found on aftermarket ECUs is built in transmission control; it can control a GM 4L60E or 4L80E in addition to the engine if you order it with this feature. Other outputs include several stages of boost control, lock up torque converters, reverse lockout, fan and fuel pump control, and one PWM and two on/off general-purpose outputs. BigStuff3 also offers traction control and internal data logging as extra options.

The Gen 3 can handle ignitions from distributor-based to coil-on-plug. Its coil-on-plug capabilities make it popular with the late model domestic V-8 crowd, especially since it can use the factory sensors on Ford modular motors and the GM LSx series. The ECU can control coil-on-plug ignitions on other engines by swapping on a 24-toothed crank trigger.

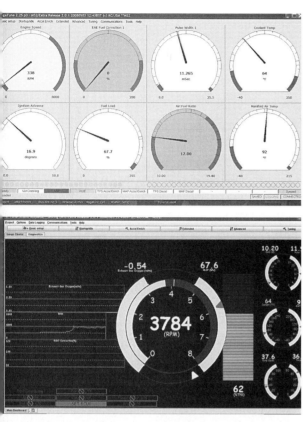

The open-source nature of Bowling & Grippo's code has encouraged several different people to write tuning software for it. These screen shots show several examples. From top to bottom, we have MegaTune, TunerStudio, and MegaTunix, which as the name implies was originally written for Linux.

They kept this through the V2.2, V3.0, and V3.57 versions of the mainboard using the MegaSquirt-I and MegaSquirt-II processors.

Given the flexibility of this system, you probably won't be surprised to hear that it's possible to configure a MegaSquirt to work with many types of factory ignition systems and sensors, and that both

distributors and hobbyists have found ways to plug it into factory wiring harnesses of many different vehicles. In the right hands it's one of the most flexible systems available with very few requirements on the sensors used, and many different ignition trigger wheel patterns supported. Plug-and-play adapters run the gamut from professionally built, polished boxes that use a MegaSquirt board inside, to homemade adapter harnesses using connectors hacked out of junkyard ECUs.

More recently, Bowling & Grippo has also begun to offer turnkey systems, fully assembled and with a more standardized hardware feature set. This includes the MegaSquirt Sequencer (an eight-channel sequential fuel and ignition EFI system with individual cylinder trim) and the MicroSquirt (a miniature water-resistant ECU for motorcycles and off-road vehicles). These keep the same tradition of making the source code readily available, even for the ready-made ECUs, so while the hardware is standardized, the firmware is being developed further by the community to add new capabilities to existing hardware. While at the time this was written MegaSquirt was traditionally batch fire, chances are the MegaSquirt-III is now the standard, with eight-cylinder sequential injection and ignition, individual cylinder trim, on-board data logging, and several other features. The emphasis with the MegaSquirt-III will still be on keeping the ECU as affordable as possible and on educating new EFIers in the process. And yes, both of the authors of this book are a part of DIYAutoTune.com, the leading MegaSquirt distributor. Just figured we'd mention that in the interest of full disclosure.

Edelbrock

Edelbrock has been in the hot rod business for 70 years. Given their experience with manifolds and carburetors, going into fuel injection is a natural extension of their products. At the moment, they offer several types of EFI systems, with the electronics outsourced. The Pro-Flow 2 and Pro-Flow XT systems are from EFI Technology, while the Pro-Tuner is made by MotoTron, a company that builds OEM engine management as well. Both are sequential systems that use speed density or alpha-N fueling, with the Pro-Tuner also offering support for mass airflow sensors or even two MAF sensors. Both can control a distributor and have internal data logging, as well. While the Pro-Tuner has mass airflow support, the Pro-Flow XT has 4 stage nitrous control, boost control, and an option of using the LS1 distributorless ignition. The Pro-Flow systems also give you the option of tuning

EFI Technologies systems often use mil-spec connectors for maximum reliability, instead of ordinary automotive grade parts.

You can buy the Edelbrock Pro-Flow system as a complete package like the one shown here.

Edelbrock builds quite a variety of intake parts. They have adapters to put many common throttle bodies onto a four barrel carburetor flange.

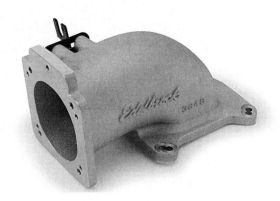

with a small handheld tuner device, although the table resolution is much smaller with the controller than with a laptop.

Naturally, Edelbrock also makes intake hardware. They sell EFI intake manifolds for many popular domestic V-8s, throttle bodies, and cast elbow adapters for putting a single-barrel throttle body on a four-barrel intake manifold. Some of their EFI manifolds and throttle bodies only come with their own engine management, but the all-out racing parts are generally available on their own. They also sell a data acquisition system called the Qwikdata 2, which is also an EFI Technology product.

EFI Technology

While EFI Technology builds the electronics for Edelbrock, which is often the sort of thing you'd see in a drag car or street rod, the sort of systems they sell under their own name are more like what you'd find at the Indy 500. While most of their products are the sort of thing you'd only see professional race teams running, they do offer ECUs in the under $2,000 range in their own

name, so these are not completely out of reach of the hobbyist, either.

The standard features make a long list, even on their entry-level models. Their Race 1.2 control unit, for example, is a four-cylinder version that allows coil-on-plug ignition control, sequential fueling with individual cylinder trim, progressive nitrous and boost control, with plenty of other inputs and outputs left over. It can control both fairly mundane devices like stepper motors and tachometers, and more exotic things like an air shifter. And that is one of their less expensive models; spending more can get you control over 12 injectors and 12 coils, mil-spec connectors, EGT monitoring, traction control, variable valve timing control, and just about anything your budget and the racing class rules can allow. They also build ignition-only control devices and multi-channel CDI systems.

Electromotive

Mention Electromotive to an experienced tuner, and often the first thing that comes to mind is their distributorless ignition systems. Electromotive got their start developing and patenting systems for precise timing control and coil charging adjustment circuits. All their current systems use a wasted spark ignition with a 60–2 crank trigger wheel. While this does prevent the use of factory trigger wheels except on engines with Bosch or similar ignition systems that already have such a wheel, this trigger wheel is extremely accurate. Sequential injection also requires a camshaft position sensor.

Electromotive's complete engine management systems are known as total engine control (TEC). These can use speed density, alpha-N, or a blend of both for their load calculation, with 16 x 16 table

Electromotive systems always use a 60–2 crank wheel. They offer kits to install this sort of trigger wheel on engines that did not originally use this sort of trigger arrangement.

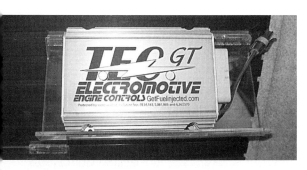

The TEC GT is Electromotive's entry-level ECU.

Electromotive system on a Cobra. Instead of a distributor, there are a set of coil packs on the firewall to fire the spark plugs.

FAST's lineup includes standalone ECUs, wideband systems, throttle bodies, and intake manifolds.

sizes. Sequential fuel control is available for up to eight peak-and-hold injectors depending on the model. Idle control options include stepper and PWM. Other features include boost control, nitrous control, internal data logging, and support for odd fire engines. There are several on/off and PWM outputs, with the exact number depending on the model.

Electromotive builds several products besides fuel-injection controllers. As you might expect, they offer bolt-on trigger wheel kits for many popular applications. They also make ignition-only control devices and develop technology for licensing to auto manufacturers. One of their recent patents covers a type of gas-electric hybrid car.

FAST

Fuel Air Spark Technology (FAST) is a division of Comp Cams that build fuel injection products with a focus on the American V-8 market. Their current XFI system is a sequential controller that supports distributor based ignitions as well as Ford modular motor, LSx, and Chrysler Hemi distributorless ignitions. The distributorless ignition control is an external module that can also work on its own without the fuel controller. Like most systems, you can choose speed density or alpha-N fuel metering. Extra features include idle air control, fan control, open loop boost control, knock sensor input, and nitrous control. It also has eight extra analog inputs and four general purpose on/off outputs, although each of the general purpose outputs is tied to specific parameters. Internal data logging is available as an add-on option.

FAST offers a wide variety of accessories with their systems, too. They sell several types of wideband oxygen sensor systems, a touch screen digital dash, and a standalone transmission controller that works with many GM, Ford, and Chrysler automatics. One of their more unusual products is the eDIST, a module that uses a normal distributor triggering signal and a one-tooth camshaft position sensor to drive a distributorless ignition on an engine that did not originally have one. For those who prefer a single coil, they also make dual sync distributors for many American V-8s. FAST offers turnkey packages for converting Ford or Chevy V-8s from carbs to EFI, and engine swap harnesses for late model Hemis and LSx motors. Comp's Inglese

Haltech's Sport 1000 ECU is part of its Platinum lineup. Photo courtesy Haltech.

Holley builds several different throttle body injection systems that can bolt in place of a four-barrel carburetor. This is one of their two-barrel designs.

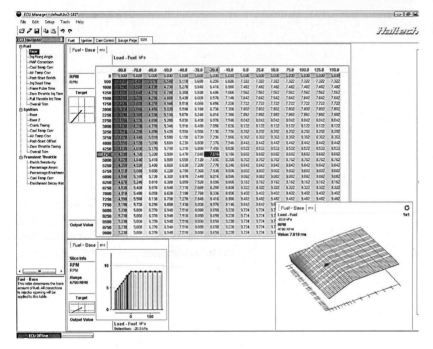

The ECU Manager software for tuning Haltech systems. Photo courtesy Haltech.

department also builds independent runner style intakes styled after Weber carburetor setups but with fuel injection, and they offer these paired with FAST's engine management.

Haltech

The Australian firm Haltech developed the first aftermarket standalone engine management system to offer real-time PC tuning. Today, they offer ECUs that span a considerable price and feature range. Most of their systems are capable of using the same outputs for fuel or ignition control. Current systems offer between five and twelve such fuel/ignition outputs, along with four PWM and two on/off outputs on most units.

Even their more basic systems have a wide range of features. All current models offer internal data logging and speed density or alpha-N fuel metering, with large table resolutions. Their ECUs support a very wide range of factory sensors, using the high-resolution signal from Nissan cam angle sensors. Some other features include boost control, torque converter control, and anti-lag and closed loop variable valve timing control on some of their models.

To simplify installation, Haltech offers adapter harnesses, which they call Plug In Patch Looms. These are available for the fourth generation Toyota Supra, second generation Mazda RX-7, and many Nissan, Subaru, and Mitsubishi applications.

Holley

Holley got their start in 1903, building carburetors. While this still represents much of their business, they made the jump to fuel injection in the 1980s with both OEM and aftermarket systems.

Their first aftermarket EFI system was the Pro-Jection, which paired a throttle body injection device with a simple alpha-N control unit that only measured rpm, throttle position, coolant temperature, and an oxygen sensor as an option. Tuning on the Pro-Jection is limited to a few knobs that the tuner adjusts with a screwdriver. The throttle body itself is similar to factory throttle bodies, but available in a four-barrel version that fits a carburetor flange.

Holley's Commander 950 ECU is sometimes confused with the Pro-Jection, as you can also order it paired with the same throttle-body injection unit.

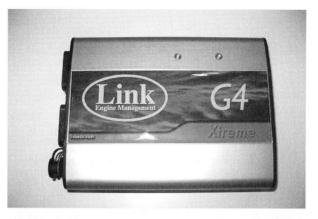

A Link G4 Xtreme ECU has a long list of inputs and outputs, including continuously variable valve timing and drive by wire throttle control.

The earliest Holley Pro-Jection ECUs, like this one, used a few knobs for adjustment. The later Commander 950 can be tuned with a laptop.

Inside a Link plug-and-play ECU. The main ECU board sits on top of a board that adapts it to the factory ECU connector.

Confusing the two would be quite a mistake, as the Commander is a very different design and a major upgrade compared to its predecessor. It is capable of speed density or alpha-N fuel metering, and lets you tune the system with a laptop. The fuel maps are 16 x 16, for over 250 more points of adjustment than the original Pro-Jection offers. The Commander also offers distributor-based ignition, cooling fan, and stepper IAC control, and can use wideband or narrowband input. It's a batch fire system with a maximum of eight injectors.

Naturally, Holley has extended their intake manifold line to include manifolds for EFI as well as carburetors. They currently offer a fairly large selection of Chevrolet intake manifolds, as well as a manifold designed for the 5.0 Mustang. Holley also builds a variety of oversized throttle bodies to replace the ones on factory EFI cars. They even offer manifold and injector power pack kits that include most of the fuel system but no ECU, for those who want to use their intake hardware but have or prefer a different brand of ECU.

Hydra

Hydra specializes in plug-and-play ECUs for Japanese imports, although they have a few systems for American and European models as well. Their Nemesis II system offers a wide range of features, including sequential injection, individual fuel trim, variable valve timing control, boost control, anti-lag, and in some cases even drive by wire throttle control. The ECU has a built-in wideband controller for NTK sensors. In some cases it allows converting cars originally equipped with a distributor to a distributorless ignition. The Hydra

system only comes as a plug-and-play system, using an adapter harness to install in place of (or sometimes in parallel to) the stock ECU.

Link

Link is a New Zealand company that offers both wire-in and plug-and-play engine management options. Their current model, the G4, comes in two versions, the lower priced Storm and the more heavily equipped Xtreme. Both of these support alpha-N, speed density, and mass airflow fuel mapping, and let you blend these methods as well. The G4 is capable of making corrections based on the fuel's temperature and pressure, and allows sequential fueling and individual cylinder tuning. The Storm has four channels of ignition output and four channels of injector output, while the Xtreme has eight of each channel. Both can work with a wide variety of factory ignition trigger wheels.

The G4 ECUs have a very long list of available outputs. They have support for anti-lag control, boost control, staged injection, practically every sort of idle air control, and many types of continuously variable valve timing. The Xtreme version also supports drive-by-wire throttles. Link also offers quite a variety of plug and play engine management systems, most of which are for Japanese cars.

Motec's newer ECUs often have unusually thin cases, but they manage to pack a lot of features into them.

At a glance, this looks like somebody machined a carburetor from billet aluminum. At least, you don't notice the lack of booster venturis. But this "carburetor" is really a Retrotek throttle body injection unit with the injectors hidden in the "float bowls."

Motec

Motec is one of the most well-known of the high-end ECU manufacturers. If you need a feature their ECUs don't have on the standard configuration list, ask, and for the right price you can probably have one built (you may need to be well funded). Motec offers systems to work with practically every factory trigger wheel setup available. In addition to common features, like boost control and nitrous control, many of the off-the-shelf systems have variable valve timing control, built in wideband control, anti-lag, and drive-by-wire support. They use 32-bit processors, which are pretty rare in the aftermarket ECU world, and most models allow separate trim tables for each cylinder rather than the simple fixed percent trim found on some less expensive sequential ECUs. They offer cylinder-by-cylinder knock sensing as an extra cost option.

It's possible to add extra input modules to a Motec, or use its data logging capabilities to show car speed and engine parameters on a race track map—or even transmit real time data from the car on the race track to the pit crew in the middle of the race, if you get one with the right options. Motec has a GPS module option for even more accurate track mapping. Their new M1 series is even scheduled to have a model that supports direct injection.

Motec has quite a range of accessories to go with their control units, as well. They offer ECU-controlled digital dashes and data logging devices. You can even order a digital dash pre-installed in a Momo or Sparco steering wheel. They also offer power distribution modules that take the place of a fuse box but have mil-spec connectors and the ability to shut down less critical systems in the event of a charging system failure. They also build plug-and-play ECUs for the Subaru WRX and Mitsubishi Lancer Evolution.

Professional Products/RetroTek

As the name RetroTek implies, one of this system's features is its old-fashioned appearance. The most visible element of most of its systems is a throttle body injection unit built to resemble a four-barrel carburetor or an old-fashioned Stromberg, although they also offer a port-injected version. Their throttle bodies hide the fuel injectors in false fuel bowls, but manage to target the injectors at the throttle blades to help the fuel vaporize.

Their other selling points are a low cost and simple installation. Some versions incorporate the ECU, injectors, and sensors into a self-contained unit that bolts to the top of the intake manifold, reducing the number of wires the installer needs to connect. RetroTek offers a returnless fuel system option as well, keeping down the number of fuel lines as well as wiring. The tuning software emphasizes using a wideband and self learning behavior to reduce the amount of tuning, particularly if you are using the system to control just the fuel. The system also gives you the option to control a Ford TFI or GM computer controlled HEI distributor, although for simplicity's sake it does not control other types of ignition.

This system uses speed density fuel metering and batch fire injector control. It can handle up to 16 high impedance or 8 low impedance injectors. Outputs besides fuel and spark include a stepper idle air control motor and support for returnless fuel systems on engines making up to 550 hp.

The two separate names sometimes cause a bit of confusion. This system is a joint venture between RetroTek Speed, who builds the ECU and does much of the design work, and Professional Products, who uses their overseas manufacturing capabilities to build much of the system's other hardware. Both companies sell the complete system under their own names.

SDS

Regardless of which EFI system you go with, Simple Digital Solutions (SDS) has a website worth checking out for their famous Tech Page. They have several interesting examples of fabricating turbo systems and fuel delivery parts, as well as miscellaneous bits of tuning advice there.

They simplify their system with a specific set of sensors included with the unit. SDS generally recommends installing magnets in the crankshaft damper to create a crank-trigger system, although they have recently come out with a system that uses an LS1's 24-tooth wheel. They offer fuel-only systems, systems that control distributor-based ignitions through an external CDI box, and a wasted spark option. Even though most SDS systems are not usually intended to work with factory ignitions, Western Motorsports has a plug-in adapter harness for Ford EEC-IV applications when used with an MSD ignition.

SDS uses speed density or alpha-N fuel metering with batch-fire injection, and can use signals from most popular wideband controllers. However, their fueling algorithm is somewhat different, as it does not use 3D tables. Instead, it uses a base-fueling amount that the user adjusts as a function of rpm, and a MAP trim number that is the same throughout the rev range. This makes the tuning essentially a pair of 2D tables instead of one 3D table, and tuning this is a somewhat different process from dialing in a normal 3D table. It is set up to drive saturated injectors, but can fire many peak and hold injectors if using a resistor box. Sensor inputs include knock input and nitrous input, while extra outputs include the fuel pump, radiator fan, a diagnostic light, and an rpm activated on/off output.

One of SDS's more unusual features is that you tune it with a handheld programming box instead of a laptop. The tuning box has some limitations, as you cannot save tunes to an external file and it only displays the contents of one load cell at a time

Professional Products also makes a variety of port-injected manifolds, including this one for a Ford modular motor.

instead of the complete map. While SDS has internal data logging, it only logs rpm, the air/fuel ratio, and the amount of fuel added, making for a somewhat more limited data log than most. On the other hand, the controller has a fairly simple interface, and it's not as bulky as a laptop. This programmer is something of a love it or hate it sort of device. In addition to the programmer, the SDS system includes a mixture control knob for mounting on the dash that can adjust fueling by 50%.

A Wide Range

That sums up our overview of some of the more popular systems on the market that you might want to start taking a look at. Many of these vendors have very informative websites with stacks of information and you can find full engine management systems from $140 on up to $10,000 or more covered here. Do your research to determine what you need. Do you need or will you actually use every last bleeding edge feature or are you looking for the best performance/price ratio?

Chapter 11
Dealing with Functions the Stock ECU Controlled

These C4 Corvette gauges have electronics controlling them, but it's not the engine management system that's wired to them. They will function whether the stock ECU is there or not.

As you know by now, ECU stands for engine control unit, and most aftermarket standalone ECUs take this in the most literal sense. They're typically limited to controlling just the engine itself and a few engine-related functions, not temperature gauges or transmissions. Sometimes the factory engineers use a broader definition of ECU and you may find that the factory ECU controlled some things that your aftermarket unit will not. Here's what to do if you find yourself in this situation.

Determine What the ECU Actually Controls—First, make sure the stock ECU really did control the function you're worried about. Cruise control, electronic gauges, air conditioning, automatic transmissions, and many other complex features appeared on cars before they had any computers onboard at all, and it's possible these may not be controlled by the factory ECU on your car. Even if you have a device that's obviously under computer control like a digital dashboard, the computer controlling it may not even be connected to the ECU. For example, the only function on a C4 Corvette's digital dash that actually has anything to do with the ECU is its average miles per gallon—the displays themselves run off a second computer with its own sensors, and do not need the factory ECU to operate.

You can often find clues about what devices the ECU actually does control by looking through the factory service manual. A device with no connection to the stock ECU obviously won't be a problem. Other times, you may discover that the device in question does connect to the ECU, but only to send information to the factory ECU, or maybe to receive a signal from the ECU that it may not absolutely need to do its job, and can continue to function without. For example, an air conditioning system may operate on its own, but the stock ECU can shut it down at full throttle.

The Parallel Installation

An installer's worst nightmare is when the factory ECU ties into a network that controls everything on the car from throttle operation to the gauges to the power windows. You'd think that the designer may have intended to make replacing the factory ECU as difficult as possible. Sometimes this may even have some truth to it.

In this case, your best option is probably what's called a *parallel installation*. You simply disconnect the factory ECU from the devices you want the standalone to control, such as the fuel injectors or the ignition coils (or both), and leave it connected to everything else controlling all the mundane stuff. You can often have the standalone share sensors with the stock ECU as long as the sensors give a voltage-based output, such as throttle position sensors. However, with

This LT1 Corvette has the engine controlled by a MegaSquirt, but the stock ECU (at right, below the coils) is still in place for functions that don't relate to the engine.

The R35 chassis Nissan GTR is a prime example of a car where the stock ECU controls devices that few aftermarket ECUs are equipped to control. When Nissan introduced the car, they claimed it would be impossible for tuners to hack into the engine management. This kicked off a race between several tuning firms to be the first to introduce an easy way to retune it, which may or may not have been Nissan's intent. Photo courtesy Haltech.

resistance-based sensors, such as most temperature sensors, you'll have an easier time getting a second set of sensors just for the standalone. If you choose, the standalone can have a completely separate network of sensors from the stock ECU.

A parallel installation lets you tune the items you need to tune, while leaving the factory ECU in control of the devices you'd rather leave exactly as they are. But sometimes the factory ECU may only control a few things that your ECU is not able to handle, or maybe for some reason you may need to remove all the factory electronics from the car. We'll first take a closer look at the things you'd need to consider when doing a parallel installation, and then we'll discuss some of the options you may have for controlling items in other ways if a parallel install isn't an option for you.

A Closer Look at "Going Parallel"

Installing an ECU in parallel amounts to disconnecting the stock ECU from what you need to tune, and connecting these things to the standalone ECU instead. You can do this with injectors, coils, IAC valves, and most anything the standalone ECU can control. Often the toughest issue is what to do about sensors. Which ones like to share? Here are some guidelines that will help out.

Throttle Position Sensor—Check the TPS signal running to the stock ECU with a volt meter. If it never goes above five volts, and if its voltage increases as the throttle opens, you can share the output of this sensor with most standalone ECUs. Just splice the standalone TPS signal wire in to the signal wire to the stock ECU. You should not try hooking the standalone ECU's 5v reference voltage

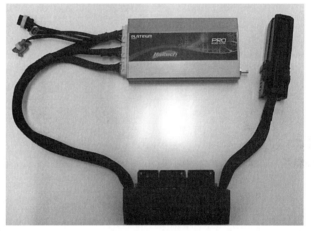

wire to the TPS, as tying two reference signals together can cause problems and the TPS already has a 5v supply.

MAP Sensor—Most aftermarket ECU's have a dedicated MAP sensor that you'll use with the new computer in addition to any stock load sensors that the factory ECU may continue to use. There's no problem with connecting two MAP sensors to a common vacuum tee. However, it is also possible to split a voltage-based MAP signal between the stock ECU and the aftermarket one. As with the TPS, don't tie the reference voltage wires together.

Temperature Sensors—These are a bit problematic. Depending on your ECU of choice you may be able to research its documentation on sharing these sensors as it is sometimes possible. If you're unsure of what to do here, your safest option is to provide the standalone ECU with its own CLT

Haltech's solution to the GTR was to build a parallel installation kit in the form of this harness and ECU. This harness connects to the stock wiring, the stock ECU, and the Haltech ECU, and disconnects the stock ECU from such functions as the injectors and ignition outputs. Photo courtesy Haltech.

and IAT sensors, separate from the stock CLT/IAT sensors that the factory ECU will continue to use.

Crankshaft Position Sensor/Camshaft Position Sensor—These are sensors you can usually share with the stock ECU, but there are a few issues that can cause the standalone to not play nice with them. If you have a VR sensor and you connect the sensor to the standalone ECU backwards, you could possibly ground the sensor and wipe out its signal. Solution: Don't wire it backwards! With Hall effect or optical sensors, sometimes the combination of pull-up resistors in the aftermarket ECU and stock ECU can interfere with each other. If this happens, one or both of the ECUs will not get an rpm signal and you may need to disable the "pull-up" inside your aftermarket ECU; check the documentation to see if this is a possibility with your ECU of choice.

Oxygen Sensor—It's perfectly okay to share narrow-band oxygen sensors with the stock ECU. However, there's a good chance you will want to add a wideband O2 sensor at some point, in order to give the standalone system more information that will be useful for tuning and closed loop control of your air/fuel ratio. These often don't work with stock ECUs, so your best bet is to weld in a second bung for the wideband sensor. This is another case where it pays to have separate sensors for the aftermarket system and the stock ECU.

So what have you really done here? Basically, your goal in the above is to make sure that both the stock ECU and the aftermarket EMS have the inputs they need from all the necessary sensors so they can both independently make all of the calculations needed to control your engine. Both systems think they are in control of the engine, as they are processing all of the incoming information from these sensor inputs, and they are sending the corresponding output signals in the way of fuel injector pulse widths and ignition pulses. So which one gets control of your engine now? Well...that's completely up to you. Which set of outputs did you actually wire up to your fuel injectors and ignition system? That's the system that's controlling your engine; the other system just continues along like an idiot thinking it's controlling the show when the programmable standalone ECU is really pulling all the strings. At least the strings that count.

What If You Get a Check Engine Light?—With older ECUs, the stock computer will often continue to be dumb enough to never notice that you cut its injector and ignition outputs. It doesn't like its sensor inputs cut, but it's not so picky about its outputs. In some cases, especially OBD-II ECUs, the stock computer might be smart enough to

figure out that you've done something tricky, which might trigger a check engine light. In some cases it may not matter, as it often wouldn't affect performance and you should only be implementing a system like this on an off-road/race-use vehicle anyway, where the CEL was of no other concern. But there are situations where this could affect performance, so you'd need to resolve the situation.

When a stock ECU activates a CEL, it often enters a limp-home mode. If the stock ECU had control of your fuel and ignition, this would generally trigger conservative fuel and ignition maps, which would reduce performance That's very likely not the issue here though, as you've got your aftermarket EMS in parallel controlling the fuel and spark. But, that stock ECU may still have control of a variable intake, variable valve timing, or throttle-by-wire system, and these could also be affected by the stock ECU entering a limp-home mode, restricting performance. So this CEL can still be a problem for you. You need to determine the cause of the CEL and trick the stock ECU into thinking everything is okay. The best way to go about this is to read the codes the stock ECU is storing to see what it's complaining about and then resolve the issue(s). Is it complaining that the injectors are disconnected? Some ECUs can't tell the difference between an injector and a 1000 ohm resistor, while pickier ones may need high wattage (50w) resistors of a similar resistance to your stock injectors.

Parallel Installation Alternatives

If you can't parallel, what options do you have? The other common alternative to a parallel installation is searching out how the designers handled a problem before putting the controls into the stock ECU. If you're lucky, you may be able to adapt systems from older models from the manufacturer, or even from other makes and models, to handle something that your ECU does. Transmissions are one of the most obvious cases, but there are other times when you may be able to pull off this sort of swap.

At other times, you may be able to get an aftermarket retrofit for older cars that don't have a certain function. Gauge kits are one common example, but with a little scavenging you can find other gadgets that may help out. For example, there are aftermarket cruise control units that you can install if the ECU you're removing used to operate on it and it's something you don't want to give up.

Air Conditioning—While turning an air conditioner on or off is pretty simple, manufacturers have come up with enough different ways to control the air conditioner that it's hard to put a

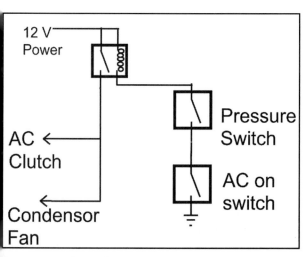

An example of an A/C control circuit with no ECU involvement. The relay is controlled through a switch on the dash and the pressure switch (which shuts down the A/C when the line pressure is too high). The relay then activates the compressor and an auxiliary fan.

One way to deal with computer-controlled gauges, particularly on a race car, is to replace the gauges at the same time you replace the stock ECU.

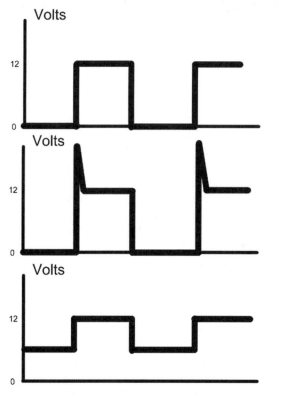

Not all tachometers use the same signal to trigger them. ECU controlled tachometers often use a 12 volt square wave (top). Tachometers that trigger off the coil may be counting on a high voltage spike (middle). And occasionally you'll see one that uses a square wave with an unusual voltage (bottom).

one size fits all circuit inside an aftermarket ECU. So unless you have a complicated digital climate control that ties into the stock ECU, you can usually make the factory air conditioning controls work by adding a few relays. The system usually follows a simple rule: Turn the air conditioning clutch on when the button is pressed, except when the pressure switch commands the air conditioning to shut down temporarily. It is fairly straightforward to rig up a circuit with relays that will behave this way. The circuit diagram at top shows one way to wire up an air conditioner where the dash switch grounds a wire to turn the A/C on and the pressure switch grounds a wire to turn the A/C off.

Many factory ECUs have a feature that shuts the air conditioner off when the engine is at full throttle. If you want to duplicate this, you can rig up a general purpose on/off output to activate a relay cutting off the A/C above a throttle position that you set.

Anti-Lock Brakes

The good news is that anti-lock brake control systems usually work on their own, without any input from the engine controls. The bad news is that if they don't, you have few options besides a parallel installation. Your best bet if you do find your ABS system is dependent on the stock ECU, and a parallel installation is for some reason not an option for you, is to look into removing the ABS system, most likely by swapping in bits from a non-ABS model of your vehicle, if available.

Gauges

The gauge the stock ECU controls most often is the tachometer. Even on cars that never had any electronics more sophisticated than an AM radio, the tachometer may expect a certain number of pulses for every rpm, and engine swaps or changing to a distributorless ignition can throw the reading off. So a fair number of aftermarket ECUs have a tach output circuit.

Unfortunately, not all factory tachometers expect the same signal. Usually an ECU's tach output is a 0 to 12 volt square wave, and this works with most aftermarket tachometers and a lot of stock designs. However, some tachometers that originally connected

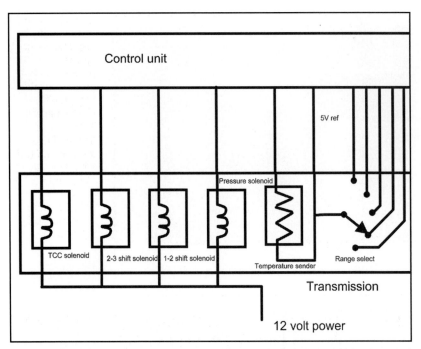

An example of the kind of output a three-speed automatic transmission might need. Some manufacturers used slightly different solenoids or switches.

to the negative terminal of the coil may be expecting a high voltage spike that this circuit doesn't provide. And there are a few oddballs out there that use signals with strange voltages, such as a 6 to 12 volt square wave. Companies like MSD offer tach adapter boxes that take a 0 to 12 volt square wave signal and convert it into something that can drive other tachometers. However, if your ECU doesn't have a tach output or it doesn't work with the tach you have, you'll need to check with the manufacturer of the tach adapter to make sure it's been tested with your gauge.

Most cars from the 1990s and earlier have all the other gauges built to run on their own, although the speedometer may share a signal with the stock ECU. But on some newer cars the engineers chose to cut down on the number of sensors and the amount of wiring by having the ECU control the gauges, either directly or by a transmission sent over a network. Sometimes, if your ECU has a user-programmable PWM output, you can configure this to drive a temperature gauge or similar device. But often, aftermarket ECUs have no provisions for gauge control beyond a tachometer. Usually, when faced with computer-controlled gauges, you'll want to use a parallel installation. However, if you need to remove the computer entirely, you may be able to swap in a gauge panel from an earlier version of your car or one with a different engine. If you don't have a junkyard choice, your only other option if

you remove the factory ECU, is likely to be a set of aftermarket gauges.

Transmission

Controlling an automatic transmission without an electronic computer requires the hydraulic equivalent of a computer, a maze of tubes, check valves, and actuators. It's no surprise that engineers started adding electronic controls to automatic transmissions at nearly the same time as they started adding them to engines.

The earliest electronic control was the lock-up torque converter solenoid. Almost any aftermarket ECU with programmable relay control outputs can take care of this. If your ECU is capable of reading vehicle speed, you can simply set the lock-up to engage once the car is moving fast enough that locking the torque converter won't stall the car. If your ECU does not have this feature, you can set it to lock up the torque converter based on rpm or a combination of rpm and throttle position.

Overdrive outputs are another type of control that some aftermarket ECUs can handle. They are usually simple on/off switches, but you'll generally need the capability to read a vehicle speed sensor and the ability to use multiple variables to control the output.

Transmissions with shift solenoids are more complicated. These need direct control from the ECU, and few aftermarket standalone systems feature automatic transmission control. There are, however, a few standalone transmission controllers available for GM and Ford transmissions. If your transmission came paired with an engine that never had computerized controls, such as a carbureted engine or some types of diesels, you may even be able to find a standalone controller in a junkyard. If no such controller exists for your transmission, and you're not able to do a parallel installation, you will either need to run a manual valve body or swap in an earlier transmission that used strictly mechanical controls.

Sometimes you'll encounter an automatic transmission with shift solenoids controlled by a separate transmission control module. If this is the case with your car, you'll need to do a little detective work to find out whether the transmission control module communicates with the electronics controlling the engine, and which way the communication goes. Sometimes the ECU may tell the transmission computer such information as throttle position and vehicle speed. Other times the transmission computer may do all the talking, telling the ECU what gear it is in and when it is shifting. Spoofing the signals a stock ECU sends to a transmission controller can be quite a challenge, but if the transmission controller only transmits information to the ECU you are on safer ground.

Throttle Control

Few aftermarket ECU manufacturers have touched drive-by-wire throttle, and with good reason. While a mistake in tuning your engine could leave you stranded on the side of the road, a mistake in tuning a throttle could send you over a guardrail. Consequently, if you're removing the stock ECU from a late-model car with drive-by-wire, or swapping a drive-by-wire engine into an older car, you may need to convert the throttle to a cable-operated setup.

If you're lucky, you may be able to find a throttle body from an older version of an engine that uses a cable and bolts right on. Or the aftermarket may offer a cable-operated throttle body specifically for your motor. Even if there isn't anything that bolts on, there's usually a fairly affordable and straightforward option. You may be able to re-drill your throttle body flange for a different throttle body design, make an adapter plate, or saw a throttle flange off a different manifold and weld it to your own. The problem is not very different from the sort you may face when converting a carbureted engine to EFI.

There is the problem of installing the throttle cable and brackets, something the engine's original designers may not have considered. If the throttle cable coils around a pulley on the throttle body, locating the throttle cable is pretty straightforward: it should come off the pulley within a few degrees of parallel to the pulley throughout the entire range of throttle travel. If your throttle cable simply attaches to a lever, you'll need to locate the throttle cable a bit more carefully. The wrong location can make the throttle touchy at low rpm or give it all its opening in the last inch of pedal travel. Having the cable at a 45-degree angle to the lever when the throttle is closed, and a 135-degree when the throttle is open, is usually a good compromise.

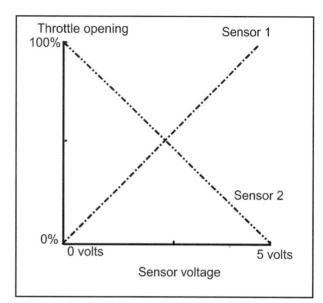

Drive-by-wire systems use multiple sensors on the accelerator pedal to avoid relying on a single sensor. These sensors usually have different output signal curves so the computer can detect if their outputs become shorted together.

Spend enough on an aftermarket standalone and you can get one that controls drive by wire, like this Link G4 Xtreme. Link's entry-level version of the G4, the Storm, does not have this feature.

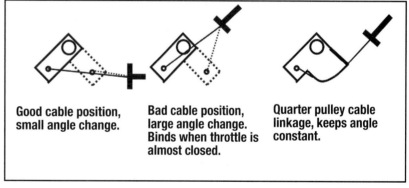

Good cable position, small angle change.

Bad cable position, large angle change. Binds when throttle is almost closed.

Quarter pulley cable linkage, keeps angle constant.

Examples of good and bad throttle cable design.

Chapter 12
Configuring and Tuning Your Engine Management System

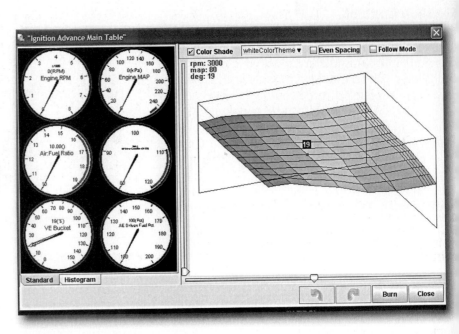

This chapter assumes you have full access to your fuel and spark tables, and can tune them in real time. If not, you will need to change the techniques a bit.

Now we get to the whole reason you've been modifying your engine management system: tuning your engine for the best possible performance. In many cases, particularly when starting from scratch with a standalone EMS, you'll first need to configure your EMS to fire the engine up. If you're lucky, you may be able to get the engine to start on the first try, but even plug-and-play systems sometimes take a bit of adjustment to get the engine to fire up that first time, particularly if the factory base map and settings have been modified. Even if the fuel injection system is up and running, getting the best power, economy, emissions and drivability out of the ECU will take a good bit of tuning. But the rewards are worth it!

We'll assume in this chapter that you have a tuning device that gives you full control of your engine, and lets you adjust your fuel and spark tables in real time while the engine is running. This could be an aftermarket standalone, a high-end piggyback, or one of the better hacks for stock ECUs. If your tuning devices don't quite reach this level, you may need to make a few adaptations based on your tuning limitations. For example, if you are adjusting a factory ECU with a chip burner that requires down time to replace the ECU's memory chip, wide-open throttle tuning won't be very different, but dialing it in at part throttle will be a lot more time consuming.

Tuning Myths

First off, let's address a few of the myths out there about standalone engine management. One is that an aftermarket ECU will not run as well as a factory unit. It's not hard to see where this one came from, as original manufacturers often have multi-million dollar budgets for developing their ECU calibrations. However, they often have to deal with considerations a race shop doesn't have to worry about. A factory ECU has to meet stringent emissions and diagnostic requirements, keep its calibration in tune even when sensors fail, or the engine racks up hundreds of thousands of miles of wear, and otherwise stand up to abuse and neglect. A race car doesn't have all of these issues, and you can get most of your drivability problems ironed out in a a few hours to a couple of days of tuning depending on how much time you spend on the details (one exception might be working on cold start issues as you need cold temps to work on these, unless you have access to a car-sized freezer like the OEMs do). And if you've modified your car to the point that the factory ECU's calibration no longer matches the engine, such as a 5.0 Mustang that has grown to over six liters and no longer has the factory cam, manifolds, or cylinder heads, the factory ECU settings won't adapt as well as an aftermarket system can. So it's really a matter of using the right tool for the job.

It's not whether you have a standalone ECU or a factory one in place that determines drivability; it's how well the ECU is tuned to match your modifications. Put a turbo on an Integra and the factory maps aren't going to be correct.

The factory ECU was well-suited for the job of running the factory engine configuration. It was safely and efficiently tuned over countless hours by many trained professionals with a limitless budget and access to tools and resources most of us don't have access to. Chances are it will run the car just as well at sea level and -40 degrees F as it will at 6,000 feet of elevation and 120-degree F temps. It will also start every time in all conditions, and will rev without hesitation. It burns the air/fuel mixture as completely as possible reducing emissions to the minimum. The stock ECU works well for what it was intended to do, but for those of you that have modified a car beyond the control of a stock computer, or who are retrofitting EFI to a car that never came with it, you need to accomplish many of these goals without the budget or resources the OEMs have.

The good news is most aftermarket standalone engine management systems, with proper tuning, can accomplish these same goals with few exceptions. They may not control some emissions equipment, for example, but can be used to make sure the mixture burns as cleanly as possible in spite of that. And you can certainly attain the drivability you desire. The cold start manners, the throttle response, smooth cruising capability, smooth transition from low load to high load at all levels of load and at all engine speeds, and of course the wide open throttle mayhem. All of this can be attained with a properly tuned aftermarket EMS. Generally, the limitation is in the tune.

If you have a naturally aspirated Saturn, copying this owner's tune might seem like a good start. But the owner has put his own custom-made camshaft position sensor under the hood (arrow). Unless you're also copying that sensor setup, you would need to adjust the settings from his tune file before attempting to start the engine.

Realistic Tuning Goals

A common mistake is obsessing over the last tenth of a point of your air/fuel ratio. While it is true that too much fuel can choke power and too little can melt pistons, having your air to fuel ratio off by 0.1 or 0.2 is not going to be a serious problem for the majority of us. It's a rare combination of sensors and wiring that can even remain this accurate outside of a lab. Although you'll need to put a good bit of attention into tuning the air/fuel ratio, many of you are reading this chapter looking for one thing: how to produce power. And the spark tables will make the biggest difference when it comes to that.

This brings up another tuning myth, the idea that the best way to tune the ignition curve is to advance it until you hear knock and then back it off. While you should back the timing off if you hear the engine knocking, this won't necessarily give you the ideal spark table. Worse, deliberately making the engine ping can put extra wear on it or in the worst of cases even cause engine failure. Knock is bad, but thankfully it is also unnecessary to ever experience knock when properly tuning an engine—in most cases anyway. More on that later.

Swapping Tuning Files—Another common stumbling block people run into (not so much a myth) is misusing other people's tuning files (the complete engine calibration from another car, saved to a file that can be shared). A lot of standalone users swap tuning files over the Internet and through car clubs. This isn't necessarily all bad, but there are some things you should watch out for. Any time you take a tuning file from an engine that is not your own, you need to exercise care. Even if it was produced by a qualified tuner, and there are plenty of maps on the Internet that definitely are not, the map probably won't be completely correct for your engine. It may just have a few details that

SETTING UP THE IGNITION

Some standalone engine controllers don't give you much choice about the ignition system you will use. Running an Electromotive TEC-3, for example, means running a 60–2 crank trigger and Electromotive's coils. There's not much that can go wrong there except getting the location of the missing tooth wrong, which their documentation will help you avoid. Systems like AEM's engine management and the MegaSquirt line of EMS systems can support an enormous variety of ignition systems. If your ECU gives you a choice of ignition systems, you will need to make sure you have made the right choice.

Some ECUs make this fairly easy if you are running a common ignition. There may be a single setting for the ignition type so you can tell it, "I'm running a GM HEI distributor," or "Use the '90–'97 Mazda Miata settings," just by going to one basic menu. Often the choice is a bit more complicated and can require a combination of different settings. This is especially true if you have something rare enough that the developers didn't include it on their standard list of ignitions. Here's are some things you will need to know to configure the ignition settings in your tuning software.

Crank/Cam Position Sensors—The first thing you will need is information about the crankshaft and/or camshaft position sensors, specifically how the triggers on these wheels are arranged. For example, if you are using the missing-tooth type trigger wheel arrangement, you will not only need to know if the wheel has 60 base teeth and 2 missing teeth, but you will need to know where the missing teeth are in relation to the sensor (in degrees) when the engine is at top dead center.

If your ignition module is not built into the ECU itself, you will need to know what sort of signal the module is expecting. Typically, the signal will go from zero to 5 volts, although some ignitions like the Ford TFI use a zero to 12-volt signal. You will also need to know if the module starts charging the coil when the voltage goes from low to high and fires the coil when the voltage goes back from high to low, or if it is the other way around. Most ignition modules follow the former pattern, letting current flow through the coil when you send them voltage and shutting off the current to produce a spark when you turn the voltage off, but there are exceptions there too. Honda ignition modules let current flow when you ground their input pin, and MSD boxes are designed to spark when their input voltage goes from low to high. Different ECUs may use different terms to tell these two types of modules apart; it may be known by such terms as "leading edge" and "trailing edge," or "inverted spark output."

The dwell time is another piece of the puzzle. Some coils need 6 milliseconds to produce their strongest spark, while others can draw so much current that applying 2.5 milliseconds of dwell time can fry ignition modules. Hotter coils use less dwell. You will need to be sure your dwell time is appropriate for your coils. If you don't know, set the dwell to a very short value. Be sure you have adjusted all of these settings to reasonable values before you connect any coils to the ECU.

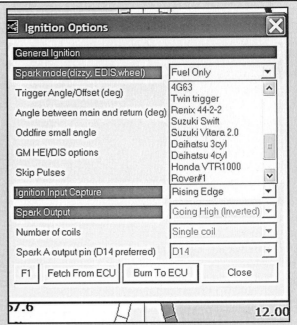

Sometimes you can select your ignition type by picking the settings from a drop down menu. Other times you will need to use numbers to represent your ignition setup using the ECU manufacturer's documentation.

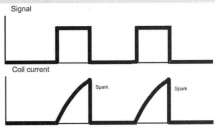

Ignition module charges on rising edge of signal, fires spark on falling edge.

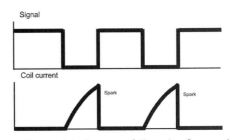

Ignition module charges on falling edge, fires spark on rising edge.

Two examples of different ignition module signals and coil current. The upper diagram shows how the ignition starts charging on the rising edge of the signal and fires the spark on the falling edge. The lower example has the opposite style input, charging the ignition when the input signal is grounded and firing the spark on the rising edge. You'll need to know which sort of module you have before you connect your coil to 12-volt power.

are different, like the size of the injectors. Even little things like the exhaust system or intake cause the breathing of the engine to be ever so slightly different, which will affect your calibration as compared to the other vehicle. Or the owner may have an engine that's just the same as your stroked Chevy LT-1, only he's changed the ignition system by installing a Bosch-style crank trigger out of disdain for the Optispark. Who cares? You might ask, "It's still the same engine, right?" Maybe not from the perspective of the computer controlling it. Unless it's 100% the same, it's not the same.

Double-Checking the Map—Bottom line, don't just load a map off the Internet and try to fire your engine up without doing some double-checking. You'll first want to look over the tune before you put it in your ECU and compare it to the documentation for your ECU to make sure the settings are correct. In particular, check to be sure the ignition settings are right for whatever ignition module you're using. Does the ignition fire off the rising or falling edge of the spark output signal? This may be called several different things by different ECUs, such as inverted spark output, leading or trailing edge, etc. You want this to be correct, though. You may even want to make the dwell settings a bit more conservative. The wrong ignition settings can, and frequently will, destroy ignition modules and can damage coils. After looking things over and checking the settings yourself, you can load it to the ECU and start tuning. But keep in mind that a tune from another car will just be a starting point. It is likely not safe to give the car full throttle on an unknown tune; in fact, it pretty much always isn't. Shared base maps should be treated as a starting point, never used as a final destination. Even start-up maps provided by the ECU's manufacturer or technical support department should be treated strictly as start-up maps. These maps are for getting the car started, not for loading onto the car and immediately cranking off a full throttle quarter mile pass, or maybe not even a test run around the corner, without tuning.

The 3D Map: Your Main Tuning Tool

The primary tuning tool on aftermarket standalone ECUs is what is known as a 3D map. You will use this as the main tool to dial in both the fuel and spark, with some help from extra tables that make corrections. 3D maps also have other applications besides fuel and spark; you may see them controlling boost or water injection, or some other outputs. A 3D table has two dimensions picked from the ECU's sensors, and a third

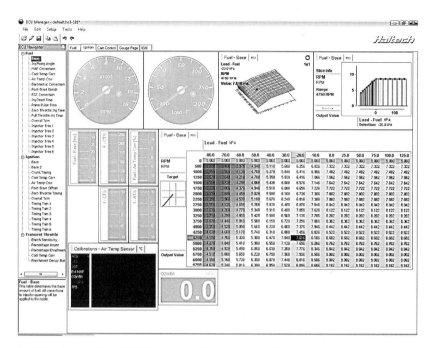

A 3D fuel map. The ECU Manager software shows the map represented as a spreadsheet, a three-dimensional contour plot, and also shows a two-dimensional slice of the map in the upper right-hand corner. This particular table is based on pulse width. Picture courtesy Haltech.

dimension is the output the ECU will give (or a figure the ECU will use to calculate that output with a little internal math). For your base fuel and ignition maps, the first two dimensions are nearly always rpm and load, where engine load can be the manifold pressure reading, mass airflow measurement, or throttle position. Most tuning software programs display 3D maps as a spreadsheet-like chart (still the 3D information, but in a flattened 2D format), usually with rpm as the horizontal axis and load as the vertical access. Most tuning programs can also display it as a three-dimensional graph, where the 3D map does indeed look like some sort of map of an island or mountain.

Reading Spark Maps—Reading the spark map is usually pretty straightforward. The number you enter into it is the spark advance at each combination of rpm and engine load, and the spark advance is usually a simple number of crank degrees.

Reading Fuel Maps—On fuel maps, the number you enter can mean different things depending on the system. If your system uses a pulse width table, the numbers represent the time, usually in milliseconds, of how long the injector turns on each time it fires. The ECU often applies other numbers to correct this for temperature or other factors, so the real injector pulse width may not match the number in the table.

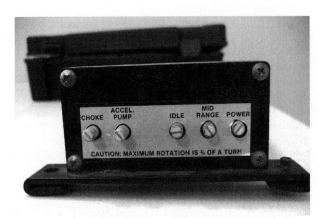

An ECU that tunes using knobs instead of a laptop. There is not very much adjustment possible on this type of system.

VE Table—The other common approach is a volumetric efficiency (VE) table. While volumetric efficiency refers to the percentage of air an engine theoretically is capable of drawing in compared to how much air it really has drawn in, the VE number on the table may not always represent the actual volumetric efficiency of the engine. Instead, the number in the VE table is a percentage that goes into a fueling equation. The fueling equation differs from ECU to ECU, but it typically starts with a base pulse width separate from the VE table. The ECU multiplies the base pulse width by the number in the VE table, then multiplies it by other numbers such as air density correction factors. Thankfully, you don't have to understand all of the math that's going on in the background here. It is accounting for a myriad of factors to apply corrections to your base fuel table and ensures that the right amount of fuel makes it into your engine when needed, once the EMS is properly tuned, that is.

Regardless of whether your ECU uses a VE table or a pulse width table for its fuel map, there's a simple rule to remember for fuel tables of most ECUS: A bigger number means to give the engine more fuel.

Fuel and spark maps come in many different sizes. Some ECUs may have an 8 x 8 map with eight rpm ranges and 8 load ranges. Others may have a 32 x 32 maps or more. This is known as table resolution and bigger is not always better. More points can give you more flexibility to deal with an engine with abrupt changes in fuel and spark needs, but they can also mean you have to spend considerably more time tuning the engine, too. Some tuners, instead of tuning every cell of a large map such as a 32 x 32 or larger map, will tune a quarter or half of the cells. Maybe tuning every other column for instance, and interpolating (or averaging) between the others instead of tuning every cell, effectively reducing the resolution of the table. Often an ECU with a smaller map will let you move the rows and columns of the map cells around in order to give you more

resolution in areas where you need it, while using just a few of the cells to cover areas where extra resolution is not needed. Some ECUs with higher resolution do not allow you to reorganize the load/rpm points in the columns/rows like this, as they have enough rows/columns that you don't really need to think about where to put them. Both methods have their strengths.

Occasionally you'll see an ECU that uses a few screwdriver adjustments to tune it. These ECUs generally use a 3D map inside, one you can't really see but instead can only make minor adjustments to with the knobs on the outside of the box. This adjustment can get a system dialed in on an engine with very similar needs to the one the designers used to create the map, but the more differences between the original engine and your own, the less likely it will be possible to fine-tune this system to perfection on your motor. Not that it won't work, but something that just works and something that's just right are very different, and in our humble opinion, one of the greatest strengths of the programmability of EFI and computer-controlled ignition is the ability to dial it in to near perfection for a given motor. You need a good bit of control to do that.

An Engine's Fuel Needs

While you can think about where to tune the fuel injectors in terms of injector pulse width, for most of us, it's probably easier to think about the engine's actual fuel needs in terms of air/fuel ratio and adjust the injector pulse width to attain that ratio. The stoichiometric air/fuel ratio for gasoline is 14.7:1, which means a mixture of 14.7 parts air

This diesel truck makes enough smoke to fill a tent, but if you can do that with a gasoline engine, you've tuned it way too rich.

to one part gasoline could burn leaving no uncombined oxygen and no unburnt fuel. Tuners often shorten "stoichiometric" to just plain "stoich." A richer mixture has more fuel, while a leaner mixture contains more air. At first glance, it seems as if 14.7:1 would be an ideal air/fuel ratio all the time. From an emissions standpoint, it frequently is the best ratio, which is why narrow band oxygen sensors are tuned to monitor the air/fuel ratio in this range. This allows the OEM engine management system to target this air/fuel ratio much of the time. However, there are several factors to consider and you need to balance several needs when fully tuning an engine, including, but not solely focusing on emissions.

Tuning for Fuel Economy Without Sacrificing Power—Fuel economy is an obvious concern. Even pure drag cars that aren't street legal may be driven at part throttle down the return lane, and there's little point in wasting valuable race gas getting back to the pit. At part throttle, an engine runs best at stoichiometric or maybe slightly leaner. Exactly how lean you can go depends on the ignition and the combustion chamber design. If you're trying to map your '57 Chevy to deal with gas prices that are far above 1957 levels, it's best to lean it out in the low load cells, but leave it richer under heavier throttle. So in the cruise areas of the map, where you'll spend time slow-speed cruising or even cruising at higher speeds but at lower loads, stoich or just lean of stoich is the way to go. On gasoline that means 14.7:1 to as lean as 15.7:1 or just a hair leaner, no further than 16:1 AFR. You'll see the best fuel economy in this range with little likelihood for introducing any reliability concerns. However, you're going to have to pay attention to your engine. Some engines may not like running much leaner than stoich and may misfire, spit, pop, or even ping as you try to lean it out, chasing fuel economy even under low-load cruise conditions. One alternate strategy if you don't mind burning a bit of extra fuel coming down that return lane, or on the road course in between hard throttle sections, is to intentionally run it a hair on the rich side to assist in cooling the cylinders. EGTs are hottest at stoich, so intentionally staying away from stoich and staying a bit richer in between the hard throttle sections can make sense on a car that spends much of its time at 10/10ths.

And, of course, we're all about the power and you'll be glad to know the above two paragraphs do not have to compromise the wide-open throttle horsepower or torque output of your engine at all. And we'd all probably agree maximum power is a good thing in a race car unless it otherwise unbalances the car. Despite the age-old adage "leaner is meaner" (which raises the obvious question...leaner than what?), a slightly richer than stoich air/fuel ratio often makes slightly more power. Generally, you get the most power in the range of 12.6:1-13.0:1 on gasoline for a naturally aspirated motor. We'll discuss proper air/fuel ratios for forced induction engines in a moment.

Whether your engine is naturally aspirated or force-fed, the proper air/fuel ratio ensures that all the air the engine draws in, reacts with fuel quickly enough and produces a fast burn time. Too little fuel will increase combustion temperatures and can lead to detonation. Too much fuel can reduce power by burning more slowly and less efficiently.

The final concern would be reliability. An overly lean air/fuel ratio can physically damage an engine by causing detonation. Adding more fuel can reduce combustion chamber temperatures and slow down combustion, helping to protect the engine from pinging or knocking. This is less of a concern at low-load part-throttle driving, where the cylinder pressures are lower and less likely to create detonation. However, at mid-high load part throttle situations (such as climbing a hill) you'll want to richen it up a bit, maybe 13.5:1-13.8:1 on gasoline. At high load wide open throttle conditions, it's going to be a range depending on the demands of the vehicle (some of these guidelines were discussed in the above paragraph). On the other side of the coin, extremely rich air/fuel ratios, such as those richer than 10:1 for gasoline, are not good for reliability either. An engine that's running extremely rich will belch soot that can clog the catalytic converter (if it has one), or what can be worse from an engine-reliability standpoint, it can allow unburned fuel to actually run down the cylinder walls, washing away the protective oil layer, leaving them without lubrication. Furthermore, as the fuel travels on down, it will dilute the rest of the engine oil, reducing the protection to the engine and allowing the engine to wear faster.

Finding the "Safe" Range—When tuning an engine you will often start out on the rich side just a bit, as it's safer to start there than to start on the lean side. But a careful tuner will pay close attention to bringing it quickly into a safe range to allow for final fine tuning. Being too rich can cause bad things to happen, although at least these consequences are not going to happen instantly like what detonation can do if you are running too lean. That safe range may be anywhere from 12:1-14:1 (ballpark) depending on what conditions the motor is being run under at the time, on the richer end under high load conditions.

Compromises

Do I need a map for power, another for economy, and another for emissions?—There are times when you need to compromise when tuning a particular map, but these compromises generally don't involve horsepower output at full throttle. Where you can usually expect to see compromise is under low-load and cruise conditions, where you'll have to choose between "best emissions with pretty good economy" or "economy is king, we should all heat our homes with tree-hugger burning stoves." Lowest emissions is going to be at 14.7:1 on gasoline, best fuel economy being closer to 15.7:1–16:1. Realistically, though, you're probably only saving 1–4 mpg depending on what beast of a motor you're feeding by trying to lean it out beyond stoich. You're probably best off sticking to more conventional methods of heating your home, and running a stoich air/fuel ratio at cruise and under low-load conditions for lowest emissions output.

Do you ever need multiple maps? Maybe. Do you run different fuels under different conditions? You might have a different map for two different grades, such as for a car that competes on both pump fuel and race fuel. Or maybe your car can run gasoline or alcohol. In these cases it's very likely you'll be changing your ignition table as well to take full advantage of that fuel.

You might have a second map if you're running a speed density–based EMS if you have any "quick mods" that you implement on your car, maybe last-minute changes as you convert from street to strip vehicle. This can be as simple and common as a cutout on the exhaust, or as far as removing the belt drive from your supercharger when running on pump gas. These changes will have a slight (or in the case of the supercharger, huge) change on the volumetric efficiency of the engine and may need to be accounted for with a second map, likely adding fuel under high loads where it's breathing better.

Different engines have slightly different needs. While the recommendations in this chapter fit many engines, some engines will like a bit more or a bit less fuel. Keep an eye on the dyno's torque output readings, as sometimes this is the only way to find out. No, we're not copping out with a "just give the engine what it wants" blanket statement here. Our intention is to try and help you understand what the your engine wants, and then it will be your job to deliver.

Determining which of the "big three,"— emissions, economy, and power—is of prime importance shifts with what load/rpm cell of the engine's fuel map you are in, and at times there is a compromise made. Special conditions such as starting, warming up, and sudden changes in rpm or throttle also have an effect on the fuel needs.

No Load Idle Tuning—Perhaps the simplest condition is when the engine is idling with no load. Here in theory it would be best to run it at 14.7:1, as this keeps emissions low and prevents the spark plugs from fouling with excess soot. However, many engines like to idle slightly richer. You just have to adjust it until the engine idles its best. You'll likely find this somewhere between 13.3:1 and 14.7:1, depending on the vehicle and how you've got your standalone EMS installed and configured. For example, sequential injection systems are more likely to be happy nearer to stoich, whereas a batch-injection system may be happier idling in the mid 13s, just like many factory batch-injection systems do. If you don't have a wideband oxygen sensor (what are you thinking?) and you're trying to get the idle about right, you can get it in the ballpark by simply tuning for the best manifold pressure reading, that is, the highest level of vacuum, as consistent as possible. This will likely be in the low 13:1 AFR range on gasoline. Regardless of where it is, your engine is happy idling here. When you do have access to a wideband, you can choose to richen/lean it out a bit if it's too far to one side or the other.

Part-Throttle Tuning—Tuning at part throttle and cruising is fairly straightforward with a wideband, somewhat more tricky with a narrowband O_2 sensor. For best emissions, you'll want to aim for 14.7:1 exactly. You'll get pretty good fuel economy here too, so this is the best all-around number to aim for at cruise/low load. As discussed previously, many engines can pick up a bit more economy by running somewhat leaner, but this also hurts emissions. Specifically, leaner mixtures tend to create more oxides of nitrogen. Different engines can be leaned out to different levels. The usual sign you've gone too far, if

"AFR Table 1"												
200.0	11.8	11.8	11.8	11.8	11.8	11.8	11.8	11.8	11.8	11.8	11.8	11.8
175.0	12.0	12.0	12.0	12.0	12.0	12.0	12.0	12.0	12.0	12.0	12.0	12.0
150.0	12.2	12.2	12.2	12.2	12.2	12.2	12.2	12.2	12.2	12.2	12.2	12.2
125.0	12.5	12.5	12.5	12.5	12.5	12.5	12.5	12.5	12.5	12.5	12.5	12.5
100.0	12.9	12.9	12.9	12.9	12.9	12.9	12.9	12.9	12.9	12.9	12.9	12.9
90.0	13.4	13.4	13.4	13.4	13.4	13.4	13.4	13.4	13.4	13.4	13.4	13.4
80.0	14.2	14.2	14.2	14.2	14.2	14.2	14.2	14.2	14.2	14.2	14.2	14.2
70.0	14.0	14.0	14.2	14.4	14.7	14.7	14.7	14.7	14.7	14.7	14.7	14.7
60.0	14.0	14.0	14.2	14.7	14.9	14.9	14.9	14.7	14.7	14.7	14.7	14.7
50.0	14.0	14.0	14.2	14.7	15.0	15.0	15.0	14.7	14.7	14.7	14.7	14.7
40.0	14.0	14.0	14.2	14.7	15.0	15.2	15.0	14.7	14.7	14.7	14.7	14.7
30.0	14.0	14.0	14.2	14.7	15.0	15.2	15.0	14.7	14.7	14.7	14.7	14.7
	500	800	1400	2200	3000	3800	4600	5400	6200	7000	7600	8200

rpm

A chart of target air/fuel ratios used on a turbocharged Miata.

Turbo engines need to run richer air/fuel ratios under boost than a naturally aspirated engine needs at full throttle.

emissions are not a concern, is if the engine starts to misfire while cruising, or loses so much power that you have to compensate by giving it more throttle and killing any economy advantage of trying to go so lean. Generally about 15.7:1–16:1 is where you'd find peak fuel efficiency.

Full-Throttle Tuning—When the engine is running flat-out at full throttle, you'll need to add more fuel. A naturally aspirated engine will typically make its best power at somewhere between 13.0:1 to 12.6:1. You'll want to test on the dyno to find out which ratio works the best on your specific engine.

Note: Don't expect massive power gains when tuning for the proper air/fuel ratio. While you can get it very wrong and wash down the cylinder walls or run it lean (both of which are wrong enough to kill power output significantly or worse), once you have the AFR close to correct, a minor change will only make a very minor change in power output. When tuning, the bulk of the power is not in the perfect AFR; fuel is tuned to get the engine running at the proper air/fuel ratio where the engine can operate safely and efficiently. You'll find the power in the ignition timing AFTER the fuel tables are right.

On turbocharged or supercharged applications, you'll want to aim for a similar air/fuel ratio as for naturally aspirated when the manifold pressure is close to atmospheric pressure. Once the boost comes up, you'll want a bit more fuel. A low-pressure turbo or supercharger, like the typical inexpensive systems designed to give a naturally aspirated motor a little more kick, say maybe 5 psi or thereabouts, typically calls for a 12.5:1 air/fuel ratio within a couple of tenths. As the boost comes up a bit, you want to head towards 12.0:1 AFR,

and engines running large amounts of boost typically need an air/fuel ratio in the vicinity of 11.8:1 to keep detonation under control. In some cases the engine may need to run even richer under extreme boost pressures, mostly using the extra fuel as a coolant in the cylinders as it's probably not all going to burn properly. At that point you might start to question if water/methanol injection isn't a better solution to cool the cylinders, as it allows you to keep your AFR closer to where it should be while providing more effective cylinder cooling than gasoline. But that's another topic for another book.

These rules assume that the engine is running at a steady throttle, whether it's cruising a flat highway or screaming down a dragstrip. They also assume the engine has started and come up to operating temperature. But you don't drive you car around all the time fully warm at constant throttle position, do you? So now we need to discuss some of the other scenarios.

Cold Start and Warm Up Fuel Needs

When the engine is starting or warming up, the cold intake manifold and combustion chambers keep the fuel from vaporizing as it should. You'll need to add more fuel the colder the engine gets in order to keep the engine running well. This varies a bit depending on your engine, intake and induction system, but there are some general rules. Typically a TBI configuration where the fuel and air mixture flows through the entire intake together will require more warmup enrichment than a multi-port configuration, with the intake carrying only air and the fuel sprayed in, near, or onto the intake valve. A lot of this comes down to the fact that with a TBI or other wet flow intake, the air/fuel mixture has more time and opportunity to allow the fuel to drop out before it reaches the combustion chamber, so you have to send more down the pipe to get a burnable mixture into the chamber to fire. Contrarily, an MPFI setup sprays the fuel in just before the intake valve and the fuel has very little time to puddle or fall out of the airstream. This results in fuel that burns much more evenly in general, and with less sensitivity to cold temps. You'll still need after-start and warm-up enrichments (WUE) either way, just more with a wet-flow arrangement. In extreme situations, the actual air/fuel ratio in a cold engine may be richer than 4.0:1, although this may not show up correctly on a wideband oxygen sensor if that fuel is puddling up somewhere, washing down the cylinder walls (which of course is bad, and is one of the reasons why cold environments are harsher on cars—they require very rich mixture to start and

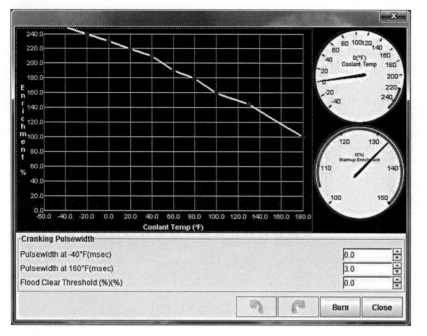

A sample warm-up enrichment curve. This one is from a small-block Chevy Nova with throttle body injection.

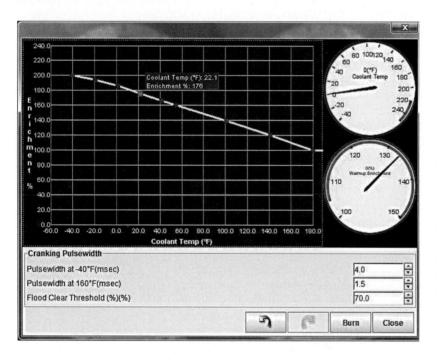

Enrichment curves are different from engine to engine. This curve is from the same Nova after the TBI was replaced with a multi-port intake.

page) is to tune your fuel and spark tables to perfection. Only after your fuel table is right can you properly tune your WUE. Once that's done, and you have the car firing up fairly easily, then tuning WUE is fairly straightforward. Since you know what air/fuel ratio you want the car to idle at, err just a hair to the rich side and adjust the WUE so that you hit this target as the car warms up. That is, let the car cool completely to whatever the ambient temperature is. Best to do this on the coolest day possible; if you do it on a warm day you may need to revisit this adjustment when the weather gets cooler. Anyways, once the engine is cool, start the car up, and while watching your wideband O2 gauge, adjust the warm-up enrichments at the current temperature to achieve the desired warm-up idle air/fuel ratio. Stay in the warm-up tool as the car warms up and hit each temperature range making sure the car is hitting the target AFR you're after.

Tip-In Acceleration Enrichments

Abrupt changes in throttle angle call for abrupt changes in injector behavior. The injectors don't just need to move to another area of the fuel map; they may need to add or subtract additional fuel because of the sudden change. Think about that last sentence for a second. Then think about this: As an abrupt change is made in throttle (maybe you jump on the gas), the computer goes from one area of the map to another very quickly. A rush of air dives into the intake, the computer needs to squirt some extra fuel into the cylinders to compensate for this rush of air until the normal load/rpm 3D tables can catch up and provide the proper amount of fuel at the new load angle. It's very similar to the accelerator pump on a carb. Something else to ponder is that the fuel the injectors spray doesn't all go straight into the cylinder; some of it ends up sticking to the walls of the intake port or manifold, and there is a time delay before it reaches the combustion chamber. When the engine is running in a steady state, the amount of fuel added to the walls is the same as the amount coming off the walls, but when the throttle suddenly opens, you need to add more fuel to compensate for the fuel sticking to the walls. Likewise, when you lift off the throttle, you can cut the fuel further because the engine will be drawing some of its fuel from the intake walls and some from the injectors.

The most basic way for the ECU to compensate for this fuel sticking to the walls is a setting that tells it to add extra fuel when the throttle suddenly opens, and possibly reduce some of the fueling when it snaps closed. Some ECUs give you the

warm up in very cold climates, not to mention frequent oil changes), or otherwise not making it to the exhaust.

So how do you get your engine where it needs to be? The first thing you need to do before bothering with WUE (maybe checking to see if the general slope is similar to the curve in the top chart on this

Black Opel Racing uses a GM 4.2 inline six on alcohol with 15:1 compression. Tuning for fuels other than gasoline follows much the same principle except you target a different air/fuel ratio.

Fuel	Stoichiometric Air/Fuel Ratio
Methanol	6.4:1
Ethanol	9.0:1
Gasoline	14.7:1
Propane	15.5:1
Natural gas	17.2:1

Tuning for Fuels Other Than Gasoline

You'll notice in this book we've given quite a few examples of air/fuel ratios, on the assumption that you're tuning for gasoline. However there are many other fuels that will burn in a spark ignition engine besides gasoline. You can run an engine on ethanol, methanol, propane, natural gas, or several other combustible liquids/gases. The biggest difference is that they have different stoichiometric air/fuel ratios (and require fuel system changes of course). To compare an engine's requirements on different fuels, engineers came up with a measurement called *lambda*. Lambda is the air/fuel ratio the engine is running divided by the stoichiometric air/fuel ratio. So for gasoline, one lambda is equal to an air/fuel ratio of 14.7. You can use lambda to compare air/fuel ratios between different fuels. If your turbo motor is running at 11.8:1 under full boost on gasoline, that's 0.80 lambda. It's likely to want 0.80 lambda on methanol or propane, too. You can use lambda to find what your air/fuel ratio needs to be on other fuels. These stoichiometric ratios can be used to convert lambda to air/fuel ratios and back. In fact, if you intend to work with different fuels frequently, it may be in your best interest not to even think in terms of air/fuel ratio when you're tuning, but in terms of lambda instead, so that you'll be able to switch between tuning for different fuels easily.

Regardless of whether you tune using lambda or air/fuel ratio as a reference, you'll still need to know the stoichiometric AFR of the fuel you're working with.

These different air/fuel ratios also mean that injector sizing is very different when you are running a fuel other than gasoline. Methanol injectors, for example, would need to be 2.3 times as large as gasoline injectors for the same engine, and if you are running high boost that calls for 11.8:1 on gasoline, you'd use an air/fuel ratio of 5.1:1.

The warm-up and acceleration enrichment settings often have major changes when switching between fuels, because the fuels have different rates of vaporization. Gaseous fuels, for example, don't stick to the intake walls. Consequently, a propane-burning engine is going to need almost no acceleration enrichment compared to an engine burning gasoline.

Hydrogen presents a special challenge. Conventional wideband oxygen sensors require hydrocarbons to operate correctly. An engine running pure hydrogen will not produce valid readings with a normal wideband oxygen sensor, making tuning a unique challenge.

option of using other tuning options that change the amount of extra fuel squirted based on engine rpm, or get more sophisticated and even attempt to internally model how the fuel sticks to the walls and predict the fuel needs based on the computerized model. This is known as X/Tau acceleration enrichment, after some of the symbols used in the equations used to model the fuel puddling.

Tuning acceleration enrichments (AE) is going to vary from EMS to EMS. Simple systems with basic accelerator pump-style enrichments basically give you a few points to adjust based on how quickly you stab the throttle. You can adjust how much fuel is squirted, typically by adding a static pulse width amount, for each throttle stab position. More modern systems using X/Tau-based modeling algorithms will

While much of the fuel drawn into the cylinder comes straight out of the injector, a part of the fuel sticks to the walls of the intake manifold. This creates a time delay that requires the injectors to add more fuel during sudden throttle openings.

often have a handful of tables to adjust.

Note that again these are settings that really should be tuned after your fuel table is properly tuned first. In fact, when tuning your fuel tables on a dyno, you should disable AE so that you don't activate AE accidentally while tuning on the dyno. This can allow the AE to influence the air/fuel ratio you're tuning for and can cause you to chase your tail a bit, particularly if the AE settings are too sensitive and allowing AE to activate when it shouldn't. Check your ECU's documentation to learn how you can disable AE on your ECU. It's often as simple as setting an activation threshold (for the speed of the minimum throttle stab that will activate AE) artificially high.

An Engine's Spark Timing Needs

Tuning spark timing is often the best kept secret in engine tuning. This is where you can often pick up very real power gains, but it's hard to safely find the power except on a proper dynamometer. While you can measure your air/fuel ratio with a wideband oxygen sensor, at this time there's no readily available and affordable equivalent sensor that can tell you how close your timing is to where it should be. Manufacturers and some well-equipped engine shops have high-speed pressure sensors that read pressure in the cylinder, but these are prohibitively expensive. Ion sensing may represent a less expensive technique for this, but it hasn't made its way to aftermarket EFI yet.

Engine timing needs are derived from the time required for the fuel to burn. Read that again if you want. (It's not that deep, but if you haven't thought through the entire process of why ignition timing is important, read that again a few times before you move on.) Because the fuel does not burn instantly, and because you want the peak of the pressure wave to occur while the piston is just on its way back down the hole on the power stroke, you will need to light the fuel mixture off before the piston reaches top dead center, giving the combustion time to begin before TDC, continue at TDC, and reach its peak pressure after TDC by about 10–16 degrees. So backing up, the cylinder pressure normally rises on the compression stroke simply due to the compression itself, but the pressure in the cylinder begins to skyrocket when the air/fuel mixture lights off, just before TDC. The pressure continues to rise as the cylinder reaches top dead center, and in a correctly tuned engine will still rise more as the piston begins to go back down, coming to a peak between 10–16 degrees after top dead center.

There are several conclusions that can be made from this analysis. One is that the ignition timing

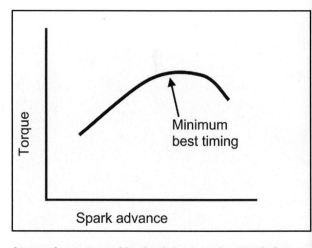

At any given rpm and load point, advancing the timing from top dead center will make more torque for a while, but then torque levels off and starts dropping. Making the cylinder pressure rise too fast can hurt power even if it doesn't lead to detonation, simply because this creates more pressure on the compression stroke.

requirements depend on how fast the fuel burns. Another is that if you light things off too early, you'll have the pressure build too fast as the piston is coming up and you'll lose power, or worse, the pressure will peak or approach peak while the piston is still on its way up towards TDC, and will try to slap the piston back down before it reaches the top. But before we go there, thinking through this you should know that advancing the timing too much can cost you power even if you don't cause detonation. The faster the fuel burns, the less timing advance you need. The engine will make the most power if the peak cylinder pressure happens 10–16 degrees after the piston reaches top dead center, exactly where in that 10–16 degree range, well, that depends a bit on the engine. So here's the scenario: The piston comes up, you've already picked the perfect time for the spark to fire (in your spark tables), and it does so, igniting the mixture. Pressures begin to rise quickly, the piston reaches TDC and starts to fall on the power stroke as the mixture continues to burn and pressures continue to rise. At somewhere between 10 to 16 degrees after TDC, the pressure reaches its peak on the power stroke and the cycle continues.

Calculating exactly when to fire the spark plug so that the cylinder pressure will hit its peak 10 to 16 degrees after top dead center is a bit too much math for your average tuner shop. When you get down to it, exactly calculating the optimal timing curve purely with mathematics is too much math for the engineers at places like Bosch, Nippondenso, and Delphi, too. While sometimes you can use a computer to predict a timing curve, there's no

While you may be able to get a decent approximation of a timing curve, finding an engine's exact timing needs will need trial, error, and careful measurement on a dyno.

If you're putting EFI on an older engine, consider starting it up on the stock ignition first, with the ECU just controlling fuel. Carrying out a project in small steps can make a complex task feel more manageable instead of feeling like it's dragging on and on.

substitute for dialing in a timing curve on a load bearing/steady-state dyno. We'll go through this process, and how to find MBT, known as Minimum Best Timing, or Maximum Brake Torque, in just a bit. But first, let's get the engine up and running.

Preflight Checklist

There are a lot of things to check on a fuel injection system before it runs. First things first: You'll want to have a fire extinguisher handy before you turn the key. It's better safe than sorry here. Even if the car didn't have to be rewired and a fuel system plumbed up from scratch, something like an intake backfire can still start a fire.

One Step at a Time—If you're new to EFI, it can greatly simplify your life to have the ECU control just one item at first, and then make it control more systems as your understanding grows. For example, if you have a previously carbureted car, you might want to have the ECU control just the fuel at first, then add ignition control later. This trick also works with many late '80s and early '90s GM and Ford products that had self-contained ignition systems; often there is a connector you can pull that will cause the ignition to ignore input from the ECU and operate on its own. (If you're using this trick you won't want to drive it like this, this typically runs the car at a static amount of timing all the time, often only 10 deg BTDC. Ever seen glowing red headers? Kind of cool looking, but not good to leave them there all the time.) If you try to take over everything at once and the car won't start, you may find yourself worrying if it is the fuel or the ignition. Of course, many installations require you to replace the stock ECU in one step and don't give you this step-by-step

option. But if you are lucky enough to be able to take control over the car in steps, it'll make troubleshooting easier and allow you to dial in your fuel table while letting the stock system control ignition. This can simplify things for you greatly if you're new to engine management systems as you'll get a hands-on education of your EMS and tuning software while working with one system, the fuel, which is probably the simpler to dial in for most people, and certainly the least intimidating. Then once that's dialed in you can take on ignition control in a separate step, putting your newfound confidence and knowledge of your EMS's tuning software to work.

Note: The following section is a rough outline on a first-start attempt with your EFI system. It's fairly generic, and you should consult your EMS documentation and use this book as a supplement only.

Check for Fuel Leaks—If you've made any modifications whatsoever to the fuel system (tank, pump, regulator, injectors, anything), the first step is to check for leaks. Turn just the fuel pump on—you may need to temporarily hot-wire the fuel pump relay to do this—and make sure there aren't any leaks. In particular, check all fuel line connections and the injectors themselves. If you have a mechanical fuel pump, you may need to unplug the ECU and crank the motor over in order to pressurize the system. If you have an adjustable fuel pressure regulator, you should also set the fuel pressure at this point.

Timing Marks—Besides a leak-free fuel system, you'll want a set of timing marks in place on the crankshaft pulley before you start the engine. If you need to troubleshoot ignition problems, you'll find them nearly impossible to figure out without a reference to a set of timing marks. Older engines are nearly always going to have a useable set of marks, although you'll want to make sure they aren't hidden under a quarter century of grime or

Before you start a new EFI setup, check to be sure the fuel system doesn't leak under pressure.

A timing marker is very useful for setting up your ECU, even if you have to improvise to put one on the engine.

After you power up the ECU, make sure all your sensors are giving reasonable readings before you start the engine. If the MAP sensor reading or temperature readings are out of whack now, correct this before you crank the motor.

affected by damage to the crankshaft pulley. On a newer engine, you may need to paint a timing mark on the pulley yourself.

Connect Your Laptop—Next, power up the ECU without cranking the engine, and connect your laptop to it. Check to be sure the ignition settings are appropriate, and if you don't know how much dwell your coil needs, set it to a conservative amount to avoid frying coils or ignition modules. You'll need to consult your EMS and ignition coil documentation for the ideal numbers here, and in some cases some test equipment may be necessary to get it just right. Sorry, we can't tell you exactly what will be a conservative enough setting for all situations here. On a very hot coil, 1 ms of dwell may be plenty and 2.5 ms of dwell could fry your ignition module, while on some ignitions, 2.5 ms would not be enough to get the engine started. It's safer to guess too low than too high. Typically, 1.5ms is a pretty safe starting point on an average coil. It should fire, but may or may not spark very hot.

Check Sensor Readings—Make sure the temperature sensors read what a cold engine should be reading and are not showing out-of-whack readings. The MAP sensor should be reading atmospheric pressure. Open the throttle fully and check if your TPS readings respond as they should. Some systems will need you to calibrate their TPS readings now, by recording the readings at closed throttle and wide open throttle. If you are using other sensors, make sure their readings are also reasonable. If one of your readings is off, check the sensor wiring and the sensor calibration in the tuning software. This is time for your all around "sanity check" with all of your systems readings if you haven't already done so. All of the sensors should be

reading reasonable numbers, they shouldn't be hopping around frantically, they shouldn't be stuck at some crazy reading, they should respond to whatever inputs they are set up to measure. Common sense goes a long way here.

Save ECU Settings—Once your sensors read as they should, save your ECU's settings. It's always good to keep a record of your settings with notes indicating what the status of your project was before you saved these.

Start Engine—Now it's time to crank over the engine and verify you are getting an rpm signal. If you're lucky (or have a good base map), the engine may start, or come close to it. Whether it tries to start or not, use this attempt to make sure you are getting an rpm signal and the rpm is reading a reasonable number in the tuning software when cranking. If it's too much to look at or you're new to your system, start a data log so you can look it over after the fact so you don't miss anything. If nothing else, you'll get some practice using your system's data logging feature, which will be priceless to you later!

If the engine does start, let it warm up, giving it whatever fuel and throttle it needs to allow it to warm up. A good quick adjustment to get the fuel close enough at idle is to watch engine vacuum and adjust the fuel to where the engine generates the highest level of vacuum. Then once warmed up, dial-in the warm idle settings and adjust the timing to smooth out the idle as much as possible. Check the timing with a timing light and make sure the timing light and the laptop agree on what your timing is. If these don't match, adjust your ECU's ignition settings and/or the appropriate sensors under the hood until the numbers match up. This

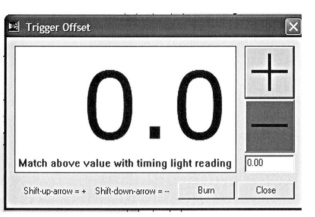

Make sure the timing you see with the timing light agrees with the timing your tuning software displays before you start tuning.

If the engine won't start, pull the spark plugs to check if they're covered with fuel or dry.

is critical; don't skip it.

If the Engine Fails to Start—If the engine didn't start, check the timing first. In this case, you'll want to unplug the injectors (or set your cranking pulse widths to something tiny like 0.01ms, writing down what they were so you can set them back). Crank the engine over while watching it with a timing light, allowing you to set the timing at cranking speed without the engine running. If you are not using the ECU to control the timing, just make sure your timing is in the right place and go on to adjusting the fuel. If your ECU is controlling both the fuel and ignition, you'll need to make sure that the spark is firing at the right time and that your ECU and the timing light agree on when the spark is firing. Adjust your ignition settings until you are getting a good crank timing and the timing reading on the laptop matches the timing light. In some cases, you may need to recheck this once the engine starts, as some ECUs use different methods for determining the cranking and running timing. GM HEI ignitions are a prime example of this: The factory computer doesn't even control timing while the engine is cranking. Many aftermarket ECUs stick to this pattern when driving an HEI ignition.

Once you've made sure you have a reasonable spark timing input on a non-running engine, it's time to plug the injectors back in and adjust the fuel settings. Your ECU manual probably covers how to estimate the fuel settings for cranking. It often helps to deliberately set the fuel added during cranking a little bit lower than you expect to need, and gradually turn it up until the engine catches. This will help avoid flooding the engine. An oxygen sensor will not provide much in the way of useful feedback for your cranking fuel needs, but examining the spark plugs and seeing if they're wet with fuel can. This can be a bit of a matter of trial-and-error, and an error that's too rich can flood the engine and make the engine nearly impossible to start until the extra fuel is gone. We've all done it—pull the plugs and dry them off, blow the cylinders out with compressed air, and start again.

If the engine fires off but dies immediately or within a few seconds, it could be that the settings for adding fuel after a cold start are wrong (likely called something to the effect of after-start enrichments and warm-up enrichments), or it could be the main fuel table is grossly wrong, or some combination of all three of these being off. Usually if it starts and then dies immediately it's a sign that there's inadequate after-start and/or warm-up enrichments, or otherwise too little fuel after the cranking process. Adjust these and try again. The goal being to try to get the engine to run, even if it's less than smooth, and nurse it along until it warms up. While it's warming up feel free to adjust the fuel table to get the idle AFR in a safe range using your wideband O_2 sensor for feedback. Then, once fully warm, adjust the hot idle areas of your main fuel and spark tables so that they are correct.

Now that the engine runs, you can move on to getting it in tune.

Tuning without a Dyno?

If you don't have access to a dynamometer, there are still a few things you can do to tune a car. There are even a few cars that make so much power they would overwhelm a dyno and have to be tuned without one, at least at full throttle. It's not that there aren't ways to do this in a very controlled environment with the right team doing the work, it's just that for the majority of us, our cars are not 3500 whp monsters. So for us there are plenty of dyno shops available with the proper equipment to do the job right.

There are two tools that you absolutely must have when tuning without a dyno: A wideband oxygen

You'll need a wideband oxygen sensor in the car if you're going to tune without a dyno.

sensor installed on the car, and data logging equipment, either built into the ECU or a laptop your passenger is carrying. A knock sensor with a warning light on the dash isn't a bad idea as a safety, but I wouldn't suggest depending on it too much.

Unfortunately it's pretty much a fact that tuning without a dyno won't get nearly as good a result as tuning with one, at least not when it comes to part throttle and drivability. You can do a fairly good job with your fuel map, but it's nearly impossible to tune the spark map properly without a dyno. You'll more than likely either leave power on the table in the name of safety, or you'll be way in the danger zone, or some combination of the two, depending on where in the map you look. But with some research and care, it is possible to get a car in the ballpark this way. In all honesty, you'll be much better off letting a qualified tuner dial your fuel and ignition maps in, but that's not what this section of the book is about. This is about giving you a best effort idea of what you can do to rough your car in on the track or other closed course...so here goes.

Find a Safe Road or Area—Many of these instructions are really for two people if you're tuning on a track. One person drives and attempts to hold the car at one rpm and load level, while the

other uses a laptop to adjust that cell of the map, then the driver moves to the next cell holding the car at a slightly different rpm or load point, and giving the tuner a chance to dial that cell in.

This really isn't something to do on public roads. Tuning often requires holding the engine at a particular throttle position for a length of time that isn't exactly convenient to hold in traffic, might tempt you to take chances that you shouldn't take or otherwise would be unsafe, not to mention dealing with the tuning equipment can be quite a distraction. This is a task that may not call for a professional driver, but it definitely calls for a closed course. A dragstrip on test and tune night is an obvious choice for working out the higher throttle area of your tables. If you have the right area to drive in while you tune, that allows you to get in to all areas of the table and tune, you can use these strategies to tune the fuel in pretty much your entire fuel table. It's a bit time-consuming to perfectly fine tune everything, but it can be done fairly well.

Determine AFR Range and Load—When tuning on a track, there are a couple indicators you'll want to watch out for. You need to have an idea of what range of AFR you're comfortable with at the load level you'll be running the engine at before you start a tuning run. For instance, maybe on a naturally aspirated vehicle you decide that 13:1 is fine at full throttle, but any leaner than 13.3:1 is leaner than you want to run it. Then you take your run watching the wideband's AFR gauge carefully, and with a data log running. If you see the air/fuel ratio gauge get too lean at full throttle, lift off the gas immediately, review your data logs, and adjust your fuel table at the load/rpm cells you were in when it began to get too lean (and dress up the lower rpm at the same level load if you want based on your logs) and then take another run. You should see your changes take effect, improving in all of the ranges you modified.

Correcting Knock—If at any time during your tuning session you hear knock, lift and hope you didn't do any damage (we did tell you we don't think this is the best way to do this, right?). Determine where you were in the map when you had the problem and correct it. Too much timing? Too little fuel? You'll also need to watch out for the usual problems like overheating, losing oil pressure, and the like while you work out your tune.

Tuning the Maps—You'll tune the fuel map first, with the spark table set fairly conservatively. You may be able to obtain a spark map from another user, or depending on what tuning solution you're using, the manufacturer may make the stock table

Even on a closed circuit with no other drivers, street tuning by yourself is not exactly a safe idea. It's much easier if one person drives while another one operates the laptop.

available. Then even if the ignition map is fine for your vehicle, we'd still recommend you pull 3–5 or more degrees out of the map. How much timing to pull depends on just how much you trust the map's creator, how similar the engine the map's creator used is to yours, and how conservative the map originally was. Sometimes it's even possible to crib (snag, snake, um... borrow) a spark map from a distributor advance curve. Again, if you pull a bad/too-advanced base map to start with, you could damage your engine, so research is prudent here. On some engines, a common advance curve may be very well known, such as the small-block Chevy, which you can rip a distributor curve from any number of websites, pull a few degrees, and start from there to be further tuned later. Keep in mind that it's possible to find or create ignition maps that are dangerous by being too conservative instead of too aggressive. Too little advance will allow still-burning fuel into the exhaust as the spark occurs late in the cycle, and combustion doesn't complete before the exhaust valves open and the exhaust stroke begins. This will cause quite a bit of heat in the exhaust, can make exhaust manifolds glow red, and in extreme cases can damage valves, turbos, or wastegates. Again, we did mention this is best done by a qualified tuner on a dyno, right?

Disabling Certain Corrections—For track tuning, some types of corrections need to be disabled when tuning the fuel map. Acceleration enrichment needs to be off, and the engine should be warmed up past the point where warm-up enrichment turns off. You have two options about what to do with oxygen sensor correction. One is to turn it off altogether. With no oxygen sensor correction, you can take the air/fuel ratio you're shooting for at a cell on the map, the air/fuel ratio you measure at that spot, and the existing reading, and calculate where you need to be mathematically. To do this, take the old cell entry, multiply it by the air/fuel ratio you measure at that combination of rpm and load, and divide it by the air/fuel ratio you want to be at in this cell. This equation sums up the process:

New cell entry = [(old cell entry) x (measured air/fuel ratio)] ÷ target air/fuel ratio

If you are using a pulse width table and your ECU does not factor in injector opening time, you'll have to modify the equation slightly by first subtracting the injector opening time from the entry in the cell, and then adding it back in at the end of the calculation. Other than that, this equation works whether your ECU goes by pulse

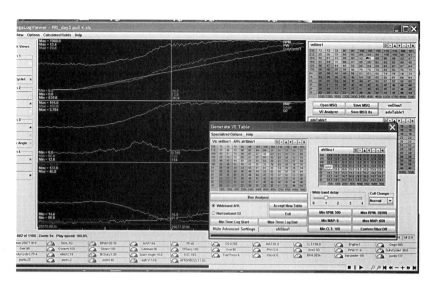

It's possible to data log the readings from a wideband oxygen sensor and use a computer analysis program to make corrections to the fuel map even without a dyno. Misfires or exhaust reversion can spoil your data, so you will need to check things by hand as well.

width or VE.

The equation may not always be perfect because of correction tables and the like, but it'll get you very close. You can then check the value you've entered by hitting that load point again. The other option, if your tuning software allows, is to enable O_2/EGO correction, set up an air/fuel ratio target table so that the O_2/EGO correction has AFR values to target at all load/rpm values, and enable a gauge to display how far the actual air/fuel ratio is from your newly setup targets while you are in each cell. Then your goal will be to tune the cell accordingly watching the gauge showing you how much EGO correction is being applied and aim for zero correction or close to it. Typically it's probably best to tune with EGO correction turned off, then you can turn it on after the tuning session, but both methods can work well.

Street Tuning Procedure—Okay, enough warnings and theory. Let's crank up the motor and start tuning. A good place to start is with the engine free revving in neutral, and tuning it for 14.7:1 A/F ratio in the map cells where it holds. Then set the adjacent, slightly higher loaded cells to similar values. If it's a pulse width table, you will want to add more fuel in these higher loaded cells. This is not necessarily the case with a VE-based table as the background math that the ECU is doing handles much of this. Keep in mind these aren't likely to be your final values as there's no load on the engine yet, but they'll be in the ballpark.

Now, tune a bit more around idle. Let the engine

sit at idle with no load and dial it in. The cells directly higher and lower in rpm are likely to be very close to the value in the cell it's currently idling, and you can generally smooth out this corner of the table (cells immediately around idle and down to the corner) with similar values to start with. Now, find ways to make the car idle differently and tune it there. That may include turning on all the electrical accessories to load up the alternator. Turn on the A/C. Put it in gear if it's an automatic, etc. You can even go so far as to manually adjust your idle speed with a throttle stop screw or air bypass screw to get it to idle in different areas. Try different combinations, hitting different cells, and dial it in getting the proper air fuel ratio. On some vehicles this will be more necessary than others, some idle along happily with little effort, others are a bit more particular. Either way you've got some ideas here on how to get it right.

Once you have the free revving and idle cells tuned, you can then go and tweak the cells the engine encounters at light cruising, driving about at part throttle. These will also be set to a 14.7:1 air/fuel ratio and give you someplace to start for more powerful part throttle (mid load/low rpm) driving, where you may choose to richen up the mixture a small amount, to the 13.8–14.3:1 range, for example. You'll work on through the table, stoich to slightly lean at cruise depending on your goals, getting richer into the 13.5 range at mid-high loads on an naturally aspirated vehicle. And then the goal will be to aim for about 12.6:1–13.0:1 at full throttle (again, naturally aspirated, boosted engines will run richer ratios on gasoline).

You can't really hold an engine steady at full throttle unless you happen to have a convenient, mile-long hill with exactly the right slope, and even then you're not going to be able to perfectly choose the cell you hold it in. So the full-throttle tuning needs to be done in short bursts, recording the air/fuel ratio as you pass through different load cells and adjusting them once the pull is over. Although you may be able to record them mentally with at least some level of accuracy, taking a data log is very helpful for this kind of tuning. Before you start, look at the lower cells and make an educated guess as to what those higher load cells should look like. The goal is to smooth out the differences from the base table you started with and your partially tuned map with the areas you haven't hit yet, conservatively estimating so you won't be too far off when you get into those cells. Err to the side of being a bit rich, as that's going to be safer than being lean for a short burst. You can often do this by looking at the trends in the table. Look at each column. If

you've tuned from 20–70kpa, for example, and you're about to tune from 70–100kpa, look at each column, starting at low rpm, and look at the change in VE or pulse width commanded in the table from 40–50kpa, then from 50–60kpa, then from 60–70kpa. These should give you a ballpark idea of how you should increase it from 70–80kpa right? Then add just a hair more for a safety net. That's the idea, look at the trends and prepare your table for what you're preparing to tune next. This will help prevent you from getting into a completely untuned and unestimated area and going completely lean or way rich. Instead, you're aiming for "just slightly on the rich side of where I want to be." Then your goal will be to quickly dial it in so you are neither too rich or too lean.

Automated Tuning Aids—Some tuning programs have the capability to use a data log for making adjustments to the fuel map. These look at the air/fuel ratio and injector pulse width, take into account corrections that were active at the time, and can analyze the data to create new fuel maps based on a data log. Such *automatic tuning* aids are handy, but you will still need to check the results yourself to be sure the map has not had errors in it. The challenge with automated tuning aids like this is they can only use the data they receive (inputs) to generate the suggested table, and that inputted data may not always be perfect.

For example, one thing to watch out for is that a rich misfire will make a fast reading wideband oxygen sensor read (falsely) lean for an instant, since the sensor reads the oxygen that did not burn with the fuel. Tuning based strictly on the oxygen sensor reading, like many automatic tuning features found in some ECU software, can respond to a situation like this and make the tune even richer and exacerbate the problem. If you are having misfires, you cannot use the oxygen sensor at that point as a reliable source of tuning. You will need to determine if the engine is too lean, too rich, or has a weak ignition problem. Solve the misfire issue, and get back to the rest of your tuning.

Track Tuning Spark Curve—It's virtually impossible to tune a spark curve at the track and get an optimal map under all load/throttle positions and rpm points. Without a high-speed way to measure cylinder pressure directly, timing needs feedback in the form of torque or acceleration measurements. You could conceivably get a decent wide-open throttle spark map by setting the timing to a fixed advance at all rpm, making a dragstrip pass and data logging, then repeating this with another fixed advance. Then review data logs to find which timing produced the largest rate of acceleration in each gear at each rpm point. This is similar to the old technique of adjusting total timing at a dragstrip test and tune night, except you can create a more detailed curve through the magic of data logging.

And yes, with enough tedious and time-consuming effort, you could potentially use this method to a good portion of the part throttle areas of the map as well, making part throttle runs, at EXACTLY the same throttle position, data logging and comparing the results. Realistically though, this technique is not exactly practical for creating an entire timing map. Trying to dial in a 12 x 12 map like this could take dozens of passes down the dragstrip at part throttle, no doubt antagonizing the track manager on those quarter-throttle pulls and 40-second ETs. About the best you can do for tuning part throttle is to adapt a rule of adding X degrees of timing for Y degrees of load (manifold pressure, MAF reading, or throttle position) below full throttle, which isn't ideal but can improve part throttle drivability somewhat. In essence, you could come up with a decent WOT curve (a very flat curve at that) and then fudge in the rest. Not pretty, definitely not optimal, but it will work if you just can't make it to a proper dyno to really do it right.

Note there are major flaws to this suggested method, as it's a compromise at best. For instance, maybe you've found that your car goes down the strip in 13.2 sec with 33 degrees of ignition advance. So you add another degree flat across all rpms at full load and you're at 34 degrees of timing. This time your ET is 13.05 seconds, an improvement. You look at your data logs and see that sure enough your rpm ramped up faster than before, so it wasn't just your mad tight shifting techniques that made the difference. You add another tick of timing and you're at 35 degrees. Try another run and you again improve your time, to 12.97. That's all fine and good, and about as good as you can expect. But if you were on a dyno, you could see that around the rpm where you make peak torque you really didn't need those last two degrees of timing and that section of the rpm range at full load is dangerously close to knocking. At the same time, other spots in the map might benefit from the additional timing, and you might even be able to use a bit more there safely for an even faster ET. Without the sort of accurate measurements you get on a dyno, it's hard to tell. The risks are real, and the gains possible on a proper dyno with a good operator are real, along with safer operation of your vehicle later.

Tuning Acceleration Enrichment—Moving on. After you have the main fuel and ignition maps

tuned, you can turn on the acceleration enrichment and tune it. This cannot be done before tuning the main fuel map, as adding the right amount of acceleration enrichment to the wrong fuel map still gives the wrong amount of fuel. The goal here is that the engine should not bog or go lean when you stab the throttle from any point in the fuel map, regardless of how hard or soft you press the pedal. However, you also should not be adding so much fuel that the engine bogs down. Tweak this until you've got enough or just a little more fuel than you need to avoid bogging or lean spikes on the AFR readout. Not that being too lean or too rich can feel very similar. Your wideband will come in handy to determine which you are, so you can adjust this parameter in the proper direction.

Remember to turn your O_2/EGO correction back on as well and adjust any related parameters as per your EMS manual. You may be able to configure an AFR target table to allow the ECU to target specific AFRs at different rpm/load points of the map, allowing you to target a fairly lean 15.3:1 economy mode cruise AFR at low-load cruise speeds, and richening up a bit to 14.7:1 or so at the same speeds but higher loads, such as when you get into the throttle a bit to climb a mild grade.

On the Dyno

Ah yes, the dynamometer. That proper tuning tool we've been telling you about. Well, some of them. First off, all dynos are NOT created equal. In the last couple of decades dynamometers have become more and more prevalent at performance shops that use them for different purposes: some for bragging rights, some to show that your car makes a few more ponies after they bolted on that shiny new exhaust system, and some that bought specialized machines that do all of the above, but have additional capabilities specifically for mapping of ECUs. These are known as *steady-state* or *load-bearing* dynos.

Inertial Dynos—So backing up, dynamometers come in two basic types: inertial and steady state. An inertial dyno uses the engine to spin a heavy drum and measures the rate of change of the drum's rpm. The original Dynojet chassis dyno fell into this category. It is good for making quick measurements of whether an engine makes more power with Part A or Part B, and for affordable basic horsepower measurements. And it's also less expensive to purchase. The disadvantage of an inertial dyno is that the engine accelerates through the entire pull simply turning a weighted drum as fast as it can accelerate it, which is not a realistic load. Additionally, you can't hold it at a steady rpm and load while examining the effects of different

A steady-state dyno is the best method of getting the entire fuel and spark mapped dialed in. They use brakes to hold the engine at constant speed. Since ordinary brakes would not stand up to this use, they use eddy current brakes like the one shown, or hydraulic pump brakes.

A roller dyno uses rollers under the drive wheels. Since this Subaru is all-wheel drive, it needs a dyno with one set of rollers for each axle.

changes you would like to make to the tables. This is not much of a liability on a carbureted motor, since very few carburetors let you make adjustments on an engine running at wide-open throttle. (In fact, Dynojet's name comes from the way the company originated, using an inertial dyno to test carb jets.) But this makes it difficult to use one of EFI's greatest tuning strengths, the ability to make changes on a running engine in real time.

Steady-State Dyno—A steady-state dyno takes a lot of the voodoo out of tuning a fuel injection system. This machine can hold an engine at a steady rpm, allowing the tuner to hold the engine at all different load points at that rpm while making changes to the mapping. The dyno displays how your changes are affecting the torque output at the wheels in real time, which is very useful information.

The steady-state dyno uses a powerful, quick-responding computer controlled brake to hold an engine at a steady rpm, or at least hold the wheels at a steady rpm in the case of a chassis dyno. It's also called a *load bearing dyno* or *brake dyno*, and the

A hub dyno uses adapters to bolt the dyno in place of the drive wheels.

brake is where the term "brake horsepower" comes from. Since ordinary brake pads would burn up in a hurry if you tried this, most steady-state chassis dynos use hydraulic or electromagnetic (eddy current) brakes. The brakes usually run under computer control, letting the operator hold the engine at a fixed rpm or allowing it to accelerate at a predetermined (and programmable) rate. Dynapack, Dyno Dynamics, and Mustang are some of the bigger players in the steady state chassis dyno game, and Dynojet now offers an eddy current brake option to some of their inertial models, which adds this capability. Although a steady-state dyno typically comes with a higher price tag, they make it easier and quicker to get a car in proper tune. Most of the advice in this chapter is for steady-state dyno tuning, since a standard inertia dyno without steady-state capability falls fairly short.

There is one other major difference between different brands of dynos: Roller dynos and hub dynos. A roller dyno puts the car's wheels on a pair of metal rollers, and uses straps to hold the car in place. A hub dyno requires unbolting the wheels and bolting a device to the hubs that transmits power to the dyno. Between these two there are minor pros and cons to each design, but you can get the job done on either one. Some argue roller dynos are a bit easier to get the car on and off of, but on the other hand hub dynos are generally safer and take traction issues out of the picture (not to mention removing the need for the four crazies in the trunk trying to get that 800 hp beast to hook up on the rollers). There are other minor pros and cons, but again, they're not showstoppers. Both designs are very capable unless you are testing a motorcycle or other vehicle that can't readily attach to a hub dyno without a special adapter.

Be sure to fix any annoying leaks before you bring the car to the dyno.

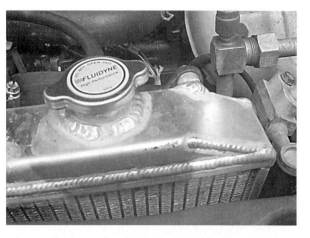

Dyno-tuning can build up a lot of heat. This Civic ruptured its plastic-tanked stock radiator on a dyno session. The aluminum radiator in there now provides more cooling and less chance of damage.

Come Prepared—Before you bring a car with a standalone engine management system to a dyno shop, you'll want to make sure you have first solved any major issues with it. There's nothing a dyno operator hates more than having to repair problems while a car is strapped down to the dyno. They'll usually express this frustration by charging you for every hour you are tying up the dyno, whether the car is running or not. And who can blame them, they could have had another paying customer on the dyno who's car didn't fall apart, right? As dyno time typically costs over a hundred dollars an hour and sometimes more than twice that amount, this can be a very expensive mistake. And if you'll be swapping out parts on the dyno, make sure to rehearse the swap beforehand and be sure your car runs in both configurations before taking it to the shop. You don't want to have to make a call to technical support that begins with, "I'm at a dyno shop so I need to hurry, but I swapped your part on and now my car won't start."

So, before you bring your car to the dyno shop, you'll want to be sure of everything in the previous section on starting the car, and make sure your car is in good repair. All the sensors should be reading correctly, the engine should be able to start up under any temperature you may encounter on the dyno, and it should idle and rev under no load without any weird ignition misfires or the like. The dwell settings should be dialed in at least well enough to keep it from eating coils or ignition modules. It should have a new set of resistor-type spark plugs, correctly gapped and in an appropriate heat range. If you've just installed a turbo or supercharger on an engine that didn't come with one, you'll probably want a smaller plug gap and a plug that's one step colder on the heat range scale than the stock plugs.

Above all, there shouldn't be any dangerous fluid leaks. You might get away with a small drip of oil, but a dyno shop will probably ask you to fix the problem and come back if you bring in a car with a fuel leak or a bad enough oil leak.

Keep an Eye on Heat—You build up a lot of heat in steady-state tuning, so if you're doing your own tuning, keep a close eye on your coolant temp sensor reading, and also your intake temps if you're running forced induction. You want to be careful not to overheat anything. In fact, if you're paying a tuner to dyno-tune your car, one of the most valuable things you can do to save money on the tuning job is to make sure your radiator and intercooler are properly sized and capable to do their jobs. The cooler the car stays, the faster and safer the tuning will go. When tuning, more time is often spent waiting on the car to cool down so tuning can continue, than doing actual tuning.

Here's a great way to think of this, especially for those of you that have added forced induction to a vehicle that didn't come from the factory that way. The motor is likely making much more power and therefore more heat. This extra heat has to go somewhere. Your cooling system had better be large

Dyno operators often use fans to cool the car. This dyno is being set up for short demonstration runs, but even in this application, the fan makes a difference.

enough to make sure that that "somewhere" isn't stuck in the engine block. You might want to consider spending a bit of money on the front end of this process upgrading your cooling system. A good radiator is an obvious investment, but depending on your car you may want a more powerful fan or other changes to the cooling system. This is a case of an ounce of prevention being worth a pound of cure—it's a pay now or pay later scenario. You can skip this and risk the dyno-tuning session taking longer due to cooling issues, or even risk blowing a head gasket. In extreme cases the tuner may have to abort the job due to cooling concerns. You're still liable for the dyno time even if your car doesn't last through the whole session, and still have to go back to the dyno when the problem is sorted out. That's going to cost you. Or, you can go ahead and spend a few bucks now on your cooling system and go to the tuner in confidence, knowing that the tuning job should go quicker and that you're prepared for the future with a solid cooling solution. Sometimes a few extra dollars for a quality aluminum radiator can pay for itself (at least partially) at the dyno shop. Be prepared and you'll have a great experience.

Tools You'll Need—There are a few tools you'll want to bring. If you're putting the car on a hub dyno like a Dynapack or Rototest, be sure to bring the right socket for your lug nuts and any sort of key if you have locking lugnuts. If you're running on a roller dyno, most shops do not allow R-compound racing slicks on the dyno. When they get hot they get soft and tend to throw chunks of rubber at high rpm. Street tires will be the better bet, or in a pinch the shop might allow a hard compound slick; ask the shop. It's probably a good idea to bring your laptop with the tuning software and whatever hardware that is needed (cable, adapter) to connect to the ECU. If the dyno shop sees a lot of cars with your ECU, they probably already have the appropriate software, but it's always a good idea to have a computer on hand that you know can connect to this ECU without any glitches. Even a shop that works with your ECU a lot may not have a program set up for your particular firmware version if you are running an unusually old or unusually new setup.

Disable Enrichments and O$_2$ Correction—Similarly to track tuning, you'll want to disable your acceleration enrichments and O$_2$/EGO correction during the base map tuning process. These don't just interfere with the tuning; they may need a few settings changes themselves once you're done with the dyno session.

The first thing to do after you put the car on a

If you have unusual lug nuts and are going to a shop with a hub dyno, don't forget to bring an adapter. The shop may not have the tools to remove specialty lug nuts.

steady-state load bearing dyno and warm it up is to setup a conservative ignition table to run while you are tuning the fuel table. As discussed earlier in the track tuning section, this can be a researched map you've dug up, the stock map, or a map the shop tuner or yourself creates from scratch if you're comfortable doing so. Bottom line: you want to aim to be 5–6 degrees retarded of what you expect to be optimum in the mid-higher load areas, giving you room to come in after you tune the fuel and dial this in.

Dial In VE/Fuel Table—Next, it's time to get the VE/fuel table perfect. You absolutely want your VE/Fuel table dialed in before you start playing with ignition advance. The spark advance table should be at conservative enough values, but if you hear any pinging (often described as sounding like BB's hitting glass, in extreme cases a light tapping of a ring on a beer bottle), lift off the gas immediately and correct the problem (generally too much timing or too little fuel). The dyno is able to hold the engine at a specific rpm, while the operator holding a steady throttle will usually keep it at the same MAP or mass airflow reading. The transmission needs to be held in a fairly high gear for steady state tuning, as this limits the torque load on the dyno. Generally you'll aim for your 1:1 ratio gear, usually fourth on a five-speed, and fifth on a six-speed. For steady-state tuning, hold it under load in each cell and dial the fuel to an appropriate air/fuel ratio for that amount of load. Typically starting at low load, low rpm while holding at that low rpm. Tune each cell, gradually increasing load up to atmospheric pressure on a naturally aspirated vehicle or just above atmospheric pressure in the case of forced

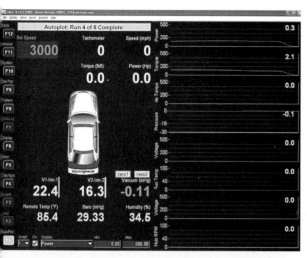

The dyno will display the engine output in real time when the wheels are turning. (In this illustration, it's showing zero as the car is stopped; the numbers are above the car outline.) Keep an eye on this number while you adjust the spark. The air/fuel readout (shown below the car) is the one to concentrate on when adjusting the fuel table; the engine output won't change very much compared to the AFR readout.

induction to maybe 70–80% of full load. Then move to the next rpm column and repeat.

Generally you'll only steady-state every cell at every load point up to the equivalent of maybe 100 mph road speed, depending on the intended use of the vehicle. A naturally aspirated car can run at full throttle even in steady-state tuning for short bursts. I'll generally rough it in up to atmospheric pressure in steady state and then confirm it and finish out the the final full-throttle tuning on the ramp runs.

Usually you do your steady state tuning holding the transmission in a 1:1 gear, but there are exceptions. This naturally aspirated rotary does not put out much torque, and we had to conduct dyno pulls in a lower gear when using it on a dyno built to handle 1,100 horsepower engines.

Just don't hold it at full throttle steady-state too long, watch your heat, and keep an ear open. Boosted cars often need to be limited to around 4 to 8 psi steady state or maybe 80% of total load, depending on how well their cooling system can handle it. You'll roll through the higher boost areas later in your ramp runs. That's the fun part that most people associate with dynos.

Repeat with Ignition Advance—Next you repeat the steady state tuning with ignition advance, except instead of targeting an AFR, you are targeting torque output directly. The idea is to hit each load cell one at a time, advancing ignition one step at a time until you don't see any more increase in torque output while in that cell. Don't add any more ignition advance than it requires to reach the maximum torque output in that cell. Then you move on to the next load cell and repeat the process. So to clarify, what the tuner will be doing is holding the engine in a particular load cell, requiring a pretty steady right foot, and adding ignition advance while watching the torque output on the dyno. Torque output will increase as you approach optimal timing, known as MBT (minimum best timing, aka maximum brake torque), then you stop there for that cell and move on to the next cell. For example, on a speed-density system reading in metric units, you might start at 40 kPa of absolute manifold pressure and 2000 rpm (basically you start as low in load/rpm as you can reach on the dyno), then staying in the same rpm column but increasing load up to the next higher row move to 50 kPa and 2000 rpm, then 60 kPa at 2000 rpm, 70kpa, 80kpa, and so on. Then after you tune all the way up to atmospheric pressure or low boost in the 2000 rpm column you move on to the next rpm column, maybe the 2600 rpm column for example, and repeat the process. Again, generally only steady state up to the equivalent of 100 mph road speed and on boosted cars up to 4 to 8 psi. You can go further if the vehicle's application calls for it, such as if it will spend a lot of time at high rpm part throttle; just know you're building a lot of heat up there and heat means risk. You'll be coming back afterward with ramp runs to tune the wide-open throttle (even on NA cars) and higher boost areas, so don't worry too much about these areas now.

Watch Your Temps—Watch the coolant temps and IAT sensor readings closely before, during, and after each tuning session. You'll have to stop at least once or twice during each column to cool down, maybe as much as three times. Any more than that and you may have cooling issues, or you could be moving too slow when tuning. A dyno session can build up a lot of heat quickly, and it's up to the tuner to make sure the temps do not get significantly above a normal range—meaning when tuning, they'll get up to the high side of normal fast—but don't let them get too high or you can cause damage. There is a lot to pay attention to when tuning. You're listening to the engine, you're controlling the dyno with a controller or keyboard, and you're controlling the car with the throttle, clutch and shifter. During pulls there are multiple temperature readings to keep and eye on, as well as air/fuel ratio readings, torque or tractive effort output readings, manifold pressure, and a myriad of other possibilities, depending on what exactly is being tuned. If you want to take the time to learn this yourself and take this on, more power to you. Take your time, read, take a class, practice and enjoy. If you take your car to a professional tuner, do the technician a favor and don't sit and ask question after question while they're trying to pay attention to all of the above. They've got a lot on their minds and they're looking to take good care of you and your car.

Tuning Problem Areas—Once you are done with the steady state tuning, you probably have a few areas where you weren't able to get with the steady state pulls. These are generally in three areas:

At higher rpm than you were willing to steady state tune the car due to heat. Often you can identify trends based on the lower columns to estimate this. Don't rely on this entirely though. Most load bearing dynos can be put into a "load mode" of some type that can add set amounts of load to the rollers/hubs and essentially simulate different grades. This can be used to run the engine into these high rpm low- mid-load areas to make sure your AFR is correct at that point without having to hold it there for long; you can just roll through the areas fairly quickly and make sure they are right. Adjust the load higher/lower and roll through at a couple different load levels to make sure you hit accurate levels of load for what you'll see in different conditions. Adjust that area of the table as needed, but unless you're going to be spending large amounts of time running the car at part throttle high rpm, you probably don't need to get carried away perfecting every cell at 8000 rpm. You can usually use the trends you can identify in the table thus far to determine where it's headed, estimate the changes, and verify your work by rolling through it under moderate load.

At higher load than you were willing to steady state tune the car due to heat (high boost). You'll catch these areas with ramp runs.

At very low load at all rpm areas. Use the same load mode mentioned earlier and experiment adjusting the load and playing with the throttle to get into the lower areas of the map to make sure you've got them dialed in. Sometimes you can get a bit lower in the load range this way, setting the dyno to very low load and rolling through at light throttle angles at different rpm. You can also try lower gears.

Preparing For WOT Pulls—The last part of the dyno tuning process is the fun part, the wide-open throttle (WOT) pulls or "ramp runs." Before you get started, look at your tables and make educated guesses based on the fully tuned cells on what the wide-open throttle cells should look like. If you're boosted, your first question may be "just how much boost will I make at full throttle?" In that case you may need to set the entire possible range up conservatively. So you'll determine this range, whether it be one or two rows near atmospheric pressure (NA car) or maybe several if you're figuring out where your boost will be, and you'll estimate the fuel and ignition needs based on lower cells you've already tuned. Follow trends you see in the table on both the X and Y axis. Then add a few percent fuel, and keep the timing several full points lower than you estimate you'll need, and you should have a good starting point.

Ramp Runs—For these pulls, you'll want to have your data logger running at any time the engine is anywhere near full throttle. Keep the files organized with names you'll be able to identify later and it can even be a good idea to keep a copy of the tune that goes with each log. You will set the dyno to the appropriate mode for ramp runs, inputting a starting and ending rpm as well as a ramp rate. The ramp rate is the amount of time, in seconds, that the dyno will allow you to accelerate from the minimum to max rpm you've set. The longer the pull, the more it loads up the engine, and the harder it is on the car. Longer pulls can sometimes build more boost as well on a turbo motor. Base this ramp rate on reality—ask yourself how long should it take your car to accelerate, in the gear you've got it in while on the dyno (fourth gear, usually), from your minimum rpm you've selected to the max. You might figure two seconds per 1000 rpm as a rough ballpark. So 10 seconds for the 1500–6500 rpm pull is a good start. If you want to go easier on the engine for the first few pulls while you settle the AFR in set it a bit shorter, get your AFR right, and then set it out to a proper length pull and confirm your AFR. You won't fully load up the engine/drivetrain getting accurate readings if your pulls are too short. If they're too long you're just putting unnecessary heat in the system.

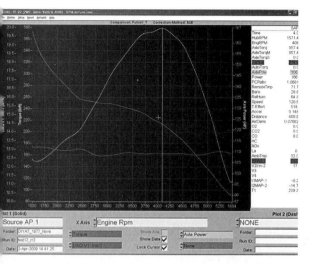

When making ramp runs, you just want to be sure your AFR readout is not heading into dangerous territory and keep your ears open for knock. Most of your adjustments will be made using the dyno's log of power, torque, and air/fuel ratio, so there is a bit less to track in real time. The unusual shape of the low rpm range of the torque curve is the effect of a torque converter.

Make a test pull at full throttle, always keeping an open ear for any knock and keeping a close eye on the air/fuel ratio. If it pings or goes lean in the slightest, lift off the throttle at once and make whatever adjustments you need to correct this problem. In theory you should have started with a conservative enough fuel and spark map so that hopefully you wouldn't be starting out with such problems...but never let your guard down. The idea with this tuning method is if you have no motor problems and adequate fuel, you'll never knock the motor while tuning, and you will get all the power you can get out of the motor. If the motor is knock limited, then you will run into knock before reaching MBT. Then you have a couple choices: solve the mechanical limitation that is causing the engine to be knock limited, or run better (higher octane) fuel. See sidebar page 124 for more details.

Making Power at WOT—So now to get started on this process of making the most power you can at maximum throttle. From here on, you're particularly looking for consistency. You want the coolant and intake air temperature readings to settle back to a similar starting temp between each pull for accurate comparisons. Let the IAT/CLT readings settle a bit, but maybe not too much. If you let it cool down a lot after the first run, you'll need to let it cool down the same amount of time between pulls for consistency. That's fine if you want, great actually, but it's more time you'll spend waiting. Generally speaking, on a car that's

perfectly happy in the 203–207 degree F range (the hot side of normal), let it settle back to the low 190s. After each pull the CLT will probably continue to climb for a few seconds, level off, and then start to come down. Let it come back down to your target CLT/IAT before the next pull. This will help ensure consistent results in your tuning, and will also remind you to keep an eye on the CLT and IAT readings during the tuning process.

Make a baseline pull and data log it. Watch your AFR readings in real time during the pull as you probably don't have WOT AFRs dialed in just yet. If it gets lean or you hear/sense any signs of detonation, get out of it and address the problem. If it stays in a safe range during the pull, let it pull on through and you've completed your baseline pull. Check your AFR readings. Check your IAT measurements and make sure they didn't get out of control; anything over about 160 degrees F is definitely something to consider addressing. Even if you're not having issues of detonation, you are definitely down on power and at higher risk of detonation than where you would be if the temps were 30 or 40 degrees F cooler. Preferably they'll max out no higher than the 130 degrees F range. Check your coolant temperature and make sure it is also in check. Review your data log to make sure both of these readings stayed in check during the pull, and check out what kind of rise you saw during the pulls, particularly with the IAT on a forced induction vehicle.

Next, look over your wideband O_2 readings in your data log. For a naturally aspirated vehicle at wide-open throttle, I'd recommend an air/fuel ratio of 12.6:1. You can experiment a bit leaner, maybe 13:1 and see if you pick up any power, but it's unlikely to make much, if any, measurable difference—12.6:1 is a good number to target on an NA motor in most situations at WOT. Spend the next few pulls on the dyno working on the fuel table smoothing out the air/fuel ratio. You won't see significant differences in power here, and that's not the goal yet. Your goal is to get as close to a solid 12.6:1 smoothly all across the pull with minimal variance and no lean spots, nice smooth AFR trace in the logs through the pull.

Forced Induction AF Ratio—If you're running forced induction, your job is the same, but the targeted air/fuel ratio is different. How different depends a bit on the car, the purpose, the turbo, and a lot of other things. We'll do our best here to give you a fairly safe starting point, but won't be able to cover all factors. To present a fairly general recommendation you might target 12.5:1 up to about 6 psi, 12:1 from 7 psi up to 12 psi, and

Knock Limited

The amount of ignition/spark advance where the engine makes peak power and the amount where it pings should not be the same. Less commonly there will be cases where attaining peak power would require more spark advance than the threshold where the engine pings/knocks. An engine like this is said to be "knock limited," in which case you have three choices. Do what we just told you *not* to do and continue tuning, bouncing it off of knock, backing it off a bit from where it was knocking, and calling it a day. That's your worst choice. You're taking the most risk tuning as you're bouncing it off the knock threshold, which will wear on the engine. Furthermore, you're relying on your ability to hear it knocking or your knock sensor to accurately sense the knock, and trusting yourself to catch it and back off in time, every time—which just does not always happen!

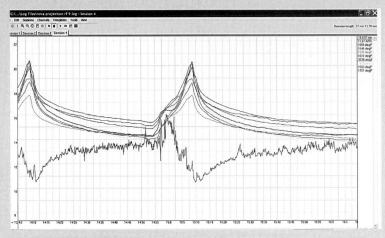

The other problem, and this should bother any true gearhead, is that you're leaving power on the table. This engine WANTS to make more power, but something is holding it back. It could be a mechanical limitation such as too hot of a heat range of spark plug, a sharp edge in a combustion chamber that's building up heat and causing pre-ignition, inefficient heat dispersal from certain combustion chambers, or an exposed piece of

Unequal fuel distribution can make an engine more prone to detonation. If your engine has this problem, you may need to tune it to run a bit on the rich side.

metal head gasket or some other foreign material in a cylinder, causing pre-ignition. Maybe your exhaust is too restrictive or your turbo too small, causing heat to back up in the cylinders. Or it could be related to an easier-to-control condition—intake temps that are too high will be much more knock prone, fuel of inadequate octane will as well. Maybe the solution is as simple as paying for premium at the pump?

If you're tuning for maximum power and torque you'll likely find you want to stick to premium fuel or better and tune for it. But what if everything here checks out and you aren't doing something crazy, like trying to run 15 psi of boost and 15:1 compression? First off, double-check to be sure you really did go through the list above of possible reasons for your knock or pre-ignition issues. If you've already done so, here are more things to check:

Make sure your AFR is spot-on where it should be (more on this later). If the engine still keeps making power when you add ignition advance right up until PING! knock city, what then? Maybe you've got an air/fuel distribution issue. You could have great air/fuel distribution to all but one of your cylinders, but have one that's getting more air than fuel, meaning it's running lean and much more knock prone. Similarly, you could have one injector that's not flowing as well as the rest, running that cylinder lean. In either case the problem cylinder will knock before the rest; that's the one you're hearing (hopefully) that stops you from being able to add the advance (that the other cylinders want) and it stops you from making the power YOU want. Not cool at all. You can either correct the root cause of the issue, try running better fuel (now and always) or leave the power on the table, essentially tuning around the problem or tuning down to the worst cylinder. These are the heartaches of a knock-limited engine.

But not all engines have this issue. Often times the amount of advance where the engine makes the best power happens before the engine gets into the region where it knocks, and advancing it to the point the engine almost or does knock will actually make less power than correctly tuning the spark table. Also there are some engines that don't have a problem, but simply need a higher grade of fuel due to the design of the engine. This is common in full race engines running high compression ratios and looking for every last bit of horsepower. An engine that is knock limited on 93-octane pump gas might not be knock limited on race gas or alcohol as these fuels may suppress detonation to a point, allowing you to reach the all important MBT ignition timing you're seeking.

11.7:1 from 13 psi on up maybe getting slightly richer. By the way, start with the boost controller turned down fairly low. You'll want to tune the low boost areas and then gradually turn up the boost a little at a time, tuning it as you go. Tune fuel at low boost, spark at the same boost level (we'll talk about ignition more in a minute), then increase boost a bit, tune fuel, spark, increase boost a bit more, and repeat until you reach your desired boost levels. That way you hit all boosted areas of the table you're likely to run the engine in at part throttle too (areas at the top and above where you steady state tuned earlier). If you find there are lower boost areas you can't get to this way due to minimum wastegate pressure, then you'll need to address this either via steady state tuning or rolling through in road mode on the dyno. Ideally you will have already steady state tuned up high enough that you can reach the remaining cells now.

Adjusting Ignition—Back to WOT tuning. Once you've gotten the wide-open throttle air/fuel ratio smoothed out and exactly where you want it, it's time to make some power, and that means playing with fire, er, timing. Add about 1 degree of timing across the range of the table you're crossing on your wide-open throttle pull. That may be straight across a single row on a naturally aspirated vehicle, or it may have a bit of a curve to it on a boosted car, unless you've set up a nice flat boost curve with an electronic boost controller (which is oh so nice, by the way).

After adding that single degree of timing, wait for the car to cool back to your baseline CLT/IAT temps, and then make a pull and see if there is a change. Check your AFRs and make sure they were still happy on that pull, there should be no significant change. Changing timing should not typically affect your AFR; you're just checking it for safety. Check your dyno horsepower and torque graph. If there is an increase all across the rpm range where you added that degree, it liked that, so try adding another. You may only see an increase in one area, in which case you'll want to pull the degree back out of the area where it didn't help. Wait for temperatures to settle back to the starting range and make another pull. The idea here is to give the engine timing where it likes it and don't give it timing where it doesn't. This process is really similar to how you earlier steady-state tuned the rest of the ignition table, only now that you're at the high load area it's not the safest thing to hold the engine at full boost and 6000 rpm. You're rolling through cells instead of holding the car in a cell while you tune.

You'll make repeated pulls, making known ignition timing changes and comparing each pull to the last. If you add a degree (or half a degree maybe) of timing and see an increase in torque output, you are either still on your way to MBT, or you might have just reached it. If you add a degree of timing and you don't see a torque increase, you very likely just took your first step past MBT at that rpm range of the map. You can try removing that degree of timing where it had no effect and make another pull to see if the torque output at that rpm range stays the same with the lesser of the two ignition timing settings, if so you've found MBT for that rpm. Once you believe you've reached MBT, you can confirm you haven't gone too far by pulling a degree of timing out and seeing if the torque drops back down. If it does, then the motor likes the timing you had there; you can put it back in. On the next pull, your confirmation should be torque coming back in at the point in the rpm range. If you find that you add a degree of timing and you actually lose torque, you've likely gone beyond MBT and you need to take that degree back out and stop at MBT. When you do that you should see the torque come back and you'll be stopping further short of the knock threshold.

Note what just happened there...you added timing a bit at a time watching torque output increase until you found MBT, you didn't know you were at MBT yet, so you tried adding another degree, and torque output either flattened out or actually decreased this time. You've entered somewhat of a buffer zone where you've added more ignition advance than will create maximum cylinder pressure, and therefore maximum torque, but you've not added so much as to create knock yet. This buffer zone may not be but a single degree of timing wide, it may be a bit wider, or in the case of a knock-limited engine, it may not exist at all. For the sake of this discussion, we'll assume you're tuning an engine that's not knock limited, and you've just added the tick of timing beyond MBT, entered the buffer zone, and torque either flattened out or dropped off a bit. What do you do? Well... since you don't need that degree, it's actually hurting power output and putting you closer to the knock threshold at the same time. Remove it, get your torque back, and the engine is dialed in right.

Keep in mind that the proper amount of advance at one area of the rev range may not hold steady through the entire pull. This is one of the benefits of electronic ignition control—the ability to customize your ignition advance curve at any rpm range—it doesn't have to just flatten out like it would with a mechanical advance distributor. So let's walk through the process of tuning your

Blindfolded Surveying

Visualizing how you find MBT can be a lot to get your mind around. So let's imagine you're not tuning an engine at all, but instead, you're a surveyor given a challenge to find the highest point on a hill. But, to make it a challenge, you have to find the highest point while blindfolded. And there's a steep cliff on the other side. At least the people who came up with this challenge let you see that the overall shape has just one peak. So when you're looking for the peak, you will need to walk carefully up the hill, one step at a time. Once you realize the hill is starting to slope back down, you're best off heading back up that hill rather than finding out just how close you can get to that edge.

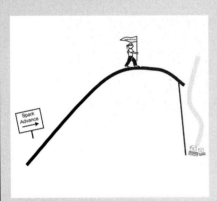

Climb the torque curve one step at a time. You don't want to walk off the edge into detonation.

ignition timing here. You start out with 26 degrees across the 100 kpa row of a naturally aspirated vehicle that revs out to 7000 rpm, you've done your research and you've found that most similar engines in similar states of tune need 31–35 degrees of advance at wide open throttle, so you figure 26 is a good starting point. Working through the process above you may find yourself raising that entire row (from maybe 2900 rpm on up) to 27 degrees and on the dyno plot from the following ramp run you see a corresponding torque increase pretty much across the entire rev range, so you leave it in and try one more. You see more torque again across the entire range with 28 degrees, so you try 29. Again, more torque everywhere, so you add another degree. When you go to 30 degrees across the entire row, maybe you don't see a torque increase below 4000 rpm, but above 4000 rpm you do. So you drop it back to 29 degrees below 4000 rpm, but then try 31 degrees above that from 4000 rpm on up. Run another test and you see no improvement below 5000 rpm, but you do see improved torque between 5000 rpm and the 7000 rpm redline. So you drop it back to 30 between 4000 and 5000 rpm where you saw no increase in torque output,

but you try 32 degrees of timing from 5000 rpm to redline. Run a test pull, check your results and see more torque from 6300 to redline but not below that. That tells you to pull the degree you just added out from 5000–6300, leaving that range at a final value of 31 degrees, but leave it at 32 degrees from 6300–7000 rpm, as it helped there. You could try for 33 degrees between 6300–7000 and see if it likes it, but at some point enough is enough—that's your call.

It's not uncommon that a motor likes to see a bit more ignition advance after the torque peak has started to drop off at higher rpms. In some cases 2–4 degrees more sloping up as the rpms climb. But you need to test it for every degree and as they say, "give it what it wants." Now you know what that means.

Once you have the ignition timing dialed in on the dyno, it's time to decide how adventurous you'd like to be. If you've found your motor to be knock limited and you've not done anything to correct the problem, then you certainly need to err to the conservative side and you'll likely leave some power on the table. If you've found that it's not, then you can likely leave your motor right there at the peak of MBT. As long as you run the same quality fuel or better, you should be in good shape.

Something else some may factor in here is the intended use of the car. A maximum effort race car may need every single pony... and the fuel being run should be adequate to make it happen, unless the rules say otherwise.

If you'd like to err on the side of caution and be more conservative, you can pull a degree or two intentionally leaving a bit of power on the table in the name of safety, since there's a chance the car may get a tank of fuel that isn't quite as detonation resistant. You're intentionally choosing to back off of MBT here, giving up a bit of power in the name of safety. You get to choose: do you want every last pony to always have fuel equal to or better than what you tuned on, leaving no room for unexpected problems such as a degrading fuel injector? Or do you want to give up just a bit in the name longevity/reliability in case of less than ideal fuel or other issues?

At WOT, you typically want your advance all in by about 2600–3300 rpm. It's fairly normal to run slightly less advance at peak torque, particularly on boosted motors, and it's not unusual at all to see a motor that's a bit happier with a little more timing at higher rpm after it has come off its torque peak. Don't just dump it in there and see what happens. Use the method described in this chapter and "give it what it wants."

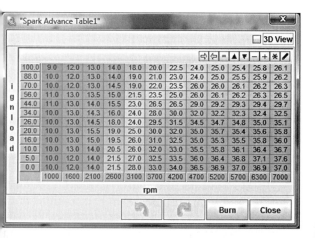

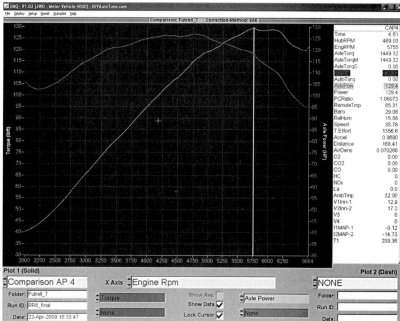

"Spark Advance Table1"

i g n l o a d	1000	1600	2100	2600	3100	3700	4200	4700	5200	5700	6300	7000
100.0	9.0	12.0	13.0	14.0	18.0	20.0	22.5	24.0	25.0	25.4	25.8	26.1
88.0	10.0	12.0	13.0	14.0	19.0	21.0	23.0	24.0	25.0	25.5	25.9	26.2
70.0	10.0	12.0	13.0	14.5	19.0	22.0	23.5	26.0	26.0	26.1	26.2	26.3
56.0	11.0	13.0	13.5	15.0	21.5	23.5	25.0	26.0	26.1	26.2	26.3	26.5
44.0	11.0	13.0	14.0	15.5	23.0	26.5	26.5	29.0	29.2	29.3	29.4	29.7
34.0	10.0	13.0	14.3	16.0	24.0	28.0	30.0	32.0	32.2	32.3	32.4	32.5
26.0	10.0	13.0	14.5	18.0	24.0	29.5	31.5	34.5	34.7	34.8	35.0	35.1
20.0	10.0	13.0	15.5	19.0	25.0	30.0	32.0	35.0	35.7	35.4	35.6	35.8
16.0	10.0	13.0	15.0	19.5	26.0	31.0	32.5	35.0	35.3	35.5	35.8	36.0
10.0	10.0	13.0	14.0	20.5	26.0	32.0	33.0	35.5	35.8	36.1	36.4	36.7
5.0	10.0	12.0	14.0	21.5	27.0	32.5	33.5	36.0	36.4	36.8	37.1	37.6
0.0	10.0	12.0	14.0	21.5	28.0	33.0	34.0	36.5	36.9	37.0	36.9	37.0

rpm

Burn   Close

The spark map for a Ford Zetec next to the dyno plot for the same motor. Note the additional spark advance after the torque peak. This particular motor is using Alpha-N to control the timing.

EFI Tuning Checklist

So to finish up here, maybe it would be helpful to list a fairly common sequence of events that a tuner might go through when tuning a car. This isn't an exhaustive list, covering things such as cold cranking pulse widths or a myriad of correction tables some ECUs have, but is meant to help with understanding the process and the sequence to use when tuning an EMS from scratch, covering all of the high spots you'll hit every time with every engine.

- Startup configuration
- Setup/Scale tables for rpm/load
- Configure ignition table on the conservative side
- Preflight checklist
- Start engine and adjust idle AFR
- Check vital signs (temps/leaks) while it warms up.
- Adjust base timing
  Disable EGO correction
- Disable acceleration enrichments
- Tune idle
- Tune no load cells near idle and low-mid rpm
- With the dyno in rpm holding mode, tune the fuel table starting at light throttle and working up to full load (N/A) in each column, working up to rpm column equivalent to a bit higher than the most common cruise speed. Do this in chunks, watching your coolant temps and intake temps. Do not let the engine overheat. For forced induction vehicles, stop short of full load, maybe at 80% of full load.
- With the dyno in rpm holding mode, Tune ignition table starting at light throttle and working up to full load (N/A) in each column, working up to rpm column equivalent to a bit higher than the most common cruise speed. Again, do this in chunks, watching your coolant temps and intake temps. Do not let the engine overheat. For boosted motors you can steady state tune higher than 100 kpa on into boost, it's just probably good to stop short of full throttle steady state under boost. Tune on up to maybe 80% of peak load.

- Estimate, looking at the trends in the tables you've built, the needs of fuel and spark for the rest of the higher rpm columns. Both fuel and spark.
- Confirm you've done this correctly. Using a load holding mode on the dyno (will apply load but not hold the engine at a specific rpm) rev the engine out at various rpms and loads. Adjust load on the dyno as you do this to get to different areas of the map and adjust air/fuel ratio as needed. You're not getting above about 80–90% load when doing this and typically paying at least as much attention to lower load areas. You can also sometimes work into some of the lower load cells this way that you may have difficulty reaching in rpm holding mode.
- Estimate what the fuel needs will be from where you stopped tuning in steady state on up to full load in all columns. It's fine to estimate a bit on the high side here. You're just roughing this in preparing for ramp run tuning of this area.
- Do the same thing for the ignition timing, but now you're going to estimate on the low side of what you expect it to be. Four or so points lower than you expect it to be is just right as a starting point.
- Ramp runs fuel: Tune the fuel at wide-open throttle. Carefully monitor the air fuel ratio and do NOT let it run lean. Listen carefully for signs of knock and get out of the loud pedal quick if

A microphone clamped onto the engine block can help you listen for detonation. Sometimes the human ear is better at signal processing than a computer.

Listening for Knock

Never let your guard down against knock when on the dyno. Keep your ears open and get off the loud pedal quick if there is any evidence. This can show up as audible knock (often described as the sound of BB's hitting glass), or it can show up as somewhat erratic torque spikes (with corresponding erratic HP spikes) on the dyno graph. That is, if the dyno you're using doesn't mask this with a massive inertia roller which will hide these spikes rolling right through them. A light-roller steady state dyno, or a hub dyno, will often show more serious knock as somewhat erratic looking spikes in the graph. Don't think if you're not seeing these you're not experiencing knock, listen closely, it's on you to identify and get out of it if you're experiencing detonation before damage is done, and it can happen very fast.

needed. Get the AFR nice and fairly smooth, right on target.

• Ramp runs spark: Tune the ignition timing at wide-open throttle. AFR should not change significantly during this. Listen carefully for signs of knock and get out of the loud pedal quick if needed.

• Enable and tune EGO Correction and any associated AFR target table

• Enable and tune acceleration enrichments

• Your engine is tuned!

EFI University

If you feel like you'd rather have some practical experience in a safe environment before you try tuning your own car, and we can understand why you would, Ben Strader and his EFI University program offer an excellent way to get a feel for what's involved in tuning a car with an experienced guru watching over your shoulder. This program consists of two courses, EFI 101 Basic and EFI Advanced, taught by experienced tuners. They're taught at various performance shops across the United States, and they sometimes take this program to other countries as well.

The EFI 101 class consists of a basic explanation of how standalone EFI works and an introduction to tuning principles. The instructor will demonstrate several tuning concepts on a car strapped to a chassis dyno, and you'll get a chance to pick up a lot of tuning wisdom from both the instructor and fellow students, who are frequently experienced at performance shops already. You'll cover most all of the tuning information that's covered in this book and then some. But more importantly you'll be able to dig in deeper into many of these subjects and ask questions of someone that's been doing it successfully for years.

In the EFI Advanced class, you'll get to run the chassis dyno yourself. This class is almost entirely hands on, as opposed to the lectures and demonstrations in the 101 class. The instructor will walk you through the basics of how the car behaves on the dyno, and you'll go through steady state fuel tuning, steady state ignition tuning, and full throttle pulls. You'll leave with enough knowledge to get started in tuning, as well as a better level of confidence in yourself tackling a tuning project, since you've been guided through it before by a pro. The classes are not cheap, but considering what you are getting they are reasonably priced, and if you plan to do a lot of standalone tuning or if you just desire a more in depth knowledge and sense of comfort before diving into your project, they are money well spent. Highly recommended. For details go to www.efi101.com.

Another option for EFI schooling that is very affordable as well as available online is Chris Macellaro's EFI Tuning Technologies online school. Take live classes with a real instructor over your PC, ask your questions, and get answers from a pro. For details go to www.efituningtech.com.

Sometimes the EFI can misbehave when it's not strictly a computer problem. A vacuum leak can be a prime example, particularly if downstream of a mass airflow meter.

Chapter 13
Troubleshooting EFI

Cars made in the 1980s and later have one more digit in the odometer than older classics, and they frequently need it. Electronic fuel injection is one of many innovations that has contributed to our modern expectation that a car is able to rack up six figures on the odometer, maybe a couple times over, but the components in a fuel-injected vehicle can and do break from time to time like anything else. This chapter covers what to do when EFI lets you down—whether something is in fact broken, or your newly installed computer just hasn't been properly trained yet on how to get your engine started.

First off, an injected engine can suffer the same mechanical breakdowns as a carbureted engine. So if your engine's leaking oil, the ECU is not at fault. Okay, maybe that example is a bit ridiculous; nobody would blame a computer for an oil leak, would they? What about a vacuum leak? What if that vacuum leak caused the computer to get an improper read on the amount of air entering the engine, and caused it to run rough, would you blame the computer as you began to get frustrated with the lack of a smooth idle? Fix the vacuum leak! Electronics have few moving parts, so before blaming the computer, you may need to rule out faults with the engine itself, such as that vacuum leak, blown head gasket, or worn rings.

So What If It Doesn't Fire Up?

Unless you do this all the time, or you somehow found a tune file from an engine that's almost identical to your

motor, having a newly installed standalone EFI system start on the first turn of the key can be the exception rather than the rule. So, let's grab a laptop and take a look at what adjustments you may need to make if the system needs a little tweaking to get the engine to fire up. By going about this in a systematic way, you can keep the hassles to a minimum. Some of the items we'll walk through really would have been considered part of the install, but they also serve as a good checklist for troubleshooting.

Check Ignition Compatibility—First, and this is a step you should do before you even connect the ECU to your ignition, you need to make sure that your ECU is configured for the type of ignition module (aka ignitor) you are using. (Of course, if your ECU is not controlling the ignition, you get to skip this step.) There are two basic types you'll usually see. One type charges the coil when it's sent a 5v signal and fires the coil when that voltage drops to zero. That's how the majority of ignitors operate. However, some ignitors operate on exactly the reverse principle; they charge the coil when the ECU grounds their input, and they fire the coil when that input signal jumps to 5v. A handful of ignition systems out there use a 12v signal instead of 5v as well.

The reason it's so important to configure your ECU for the proper type of module is the ECU only attempts to send the command to charge the coils for very brief periods of time, called the *dwell period*. Then it turns the ignitor back off, generally for periods much longer than the dwell period, at least when running at lower rpm. If you've got your ECU configured for the wrong type of module, you may still get a spark at first, but your timing will be off, and the bigger problem is that your ECU is charging your coil when it thinks it's turned off, and it's turned off when it should be charging. So not only is your timing off, but it's charging MUCH longer than it should be, which will put a lot of heat into the ignitor and the coil. Do this for any length of time and you'll usually burn one or the other up in short order, seconds in some cases. Note that this is not a configurable option on some ECUs, and most likely on those ECUs they will have a very specific list of ignitors/coils that can be used with the system. Some systems, on the other hand, are incredibly flexible, allowing nearly any ignitor or coil to be used, making it important that you configure the system properly.

Visually Check Fuel System—Another early check, if you have made changes to the fuel system when installing the standalone system, is to turn on the fuel pump and visually inspect the fuel system, usually wiring the pump to battery power so you can keep it on without the engine running. Let it run for about 10 seconds to maybe a minute or more. If it's flowing fuel, you won't overheat it, so take your time. Just be ready to disable it if you spot a leak, and keep a fire extinguisher nearby. Check the fuel pressure with a gauge, and make sure you have no leaks in the system. If you spot either problem, fix it before you go any further.

After you've made sure the ignition module is set up correctly and you have good fuel pressure, power up the ECU and connect it to your laptop or tuning tool. Bring up whatever mode displays the sensor data. See if all the readings are reasonable. The temperature sensors should read more or less the temperature of your garage unless you've been cranking the engine a while (with some starts or near starts adding heat to the system), the battery voltage should be in the healthy 12–13 volt range, the throttle position sensor should read closed, and on a speed density system, the MAP sensor should read the ambient air pressure. Any of these sensor readings being off can keep you from starting. A false cold reading could trigger too-large of a cranking pulse width and too much cold start enrichment, and flood your engine. If the ECU is set for the wrong sort of MAP or MAF sensor, its fueling calculations will be way off.

Clearing a Flooded Engine—If you do flood the engine, many ECUs have a *flood clear mode*, perhaps triggered by flooring the throttle while cranking or some other trick, which will shut down the injectors and help clear a flooded engine; check your ECU's documentation.

Make sure you have a fully charged battery, especially if the engine has been sitting for a long time. The healthier the battery, the easier the first start will be.

If all else fails, you can briefly set your cranking pulse widths very low, maybe 0.1ms, and turn the engine over like this to help clear the cylinders if the flooding isn't too bad. If you flood it too bad, pull the plugs and dry them out. (Remember to set your cranking pulse width back to normal if you used that trick). On the flip side, if your ECU does use a trick such as a wide-open TPS to trigger flood clear mode, and your throttle position sensor is giving you a false wide open reading when the throttle is closed, this may accidentally cause flood clear mode to kick in and prevent the engine from starting normally. You can see if this could be the culprit of your starting woes by looking at a data log. What are your pulse widths during cranking? If an ECU's flood clear mode is triggered, chances are you'll see zero pulse width in the log when turning the engine over.

Check Battery Voltage—If your battery voltage i low and you think you might be cranking on the engine a bit, it's time to get out the charger. Even a stock ECU that a manufacturer spent potentially millions to design and tune can have a hard time starting an engine with a spent battery. If the battery is fully charged, make sure your ECU is getting a good source of power and that it has been grounded correctly to the engine block. We'll cover how to deal with bad sensors in the second half of this chapter.

Check for Tach Signal—Now it's time for one last sensor check. Turn the key and try to crank the engine up. Watch your tuning software's interface and make sure you're getting a tachometer reading in the software (not your dash gauge), and that it's a reasonable number. This should be a nice steady

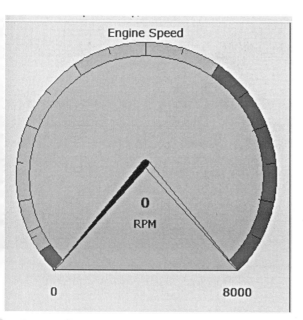

One key reading to check if your engine isn't starting is the tach readout in your tuning software. If the ECU doesn't think the engine is turning, it will not fire the injectors or spark.

If you are getting an rpm reading and the engine cranks but won't start, make sure your timing is where it needs to be. Photo courtesy James Southall.

reading, not jumping around all over the place or dropping out at all. If so, great, and you can move on. If not you'll need to investigate the source of the dropping or unstable tach signal. It could be that your crank trigger sensor is misadjusted, or that the wiring is routed too close to a source of electrical noise such as the coils or plug wires. I've even seen cases where a failing starter was causing noise on the crank position signal that made it impossible to get a clean tach signal when cranking. The armature was shorting out, causing the starter to draw massive amounts of current and to put massive amounts of noise into the entire electrical system. (A failing starter with this issue will also tend to get abnormally hot.) Replacing the starter caused the tach signal to smooth out perfectly, and the car to crank instantly. The bottom line is, if your ECU is not seeing any rpm or is not seeing a steady and noise-free signal, it will not properly fire the coils or the injectors.

Check Crank Position Sensor—The first thing to check if you are not getting rpm is to make sure your ECU is configured for your type of crank position sensor. If the rpm signal comes from anything more complicated than a basic distributor wheel, make sure the ECU is set up to read this trigger wheel or both wheels correctly. If the ECU is setup for a missing-tooth wheel and therefore is looking for a gap in the teeth that isn't there, you can crank the engine until the starter wears out and it won't fire the spark plugs. If you don't have a

tach input signal to the ECU, stop here and figure that out. You can't go any further until you do.

Check Ignition Timing—After you have verified all the sensor readings are correct, the next thing to check is ignition timing. If the timing is too far advanced or retarded, the engine has no chance of starting. If the spark is firing but the timing is off, adjust your ECU settings, distributor, or cam angle sensor to get an appropriate cranking timing, usually in the vicinity of 10 degrees BTDC. If you're able to turn the engine over at a fairly steady speed, you'll be able to put a timing light on the engine even while cranking and adjust the timing to get it in this range. Note that as a general rule, if you have too little advance when cranking, the motor may try to "pop off." If you have too much advance when cranking, the motor may "kick back" against the starter, which in some cases can even damage the starter. Somewhere in the middle is the sweet spot you're looking for.

If you're not getting a spark at all, check your ignition wiring and settings. Some vehicles such as Ford owners with a TFI or external EDIS ignition have an easy short cut here: They can pull a diagnostic connector that causes the ignition system to fire on its own, with no input from the ECU. This can let you sort out the fuel issues to get the engine started, and let you sort out the ignition later. Many other ignition systems from the early days of EFI have similar tricks that can let the ignition fire on its own at a known amount of base timing.

No Timing Marks?

Back in the days of distributors with inbuilt advance mechanisms, an engine needed marks to allow a mechanic to check and set the timing. Newer engine designs frequently determine the timing off a reluctor ring cast into the crank itself. It can be held in place with a keyway so that it's always in time with the crank. This both allows the assembly plant to set up the engine without the need to adjust the timing and makes it impossible for a mechanic to set the timing on a stock engine. Since an engine like this has no need of timing marks, designers have started leaving them off on some engines.

This is all well enough if you are using the factory ECU, or if the manufacturer of your ECU has studied this engine well enough to supply you with enough information about the sensors. You can get the timing dead-on by entering a set of numbers or a base map they provide. But if you aren't completely sure what settings you need, you will need to find a way of adding timing marks to the engine. If you're lucky, you may be able to find an older version of your engine that comes with a pulley with timing marks and a pointer you can bolt to the front of the block.

More often, however, you'll find that you need to conjure up a set of timing marks from scratch. Often, the best way to do this is to find a suitable place on the engine block where you can put a pointer mark of some sort. Then insert a piston stop into the #1 spark plug hole and rotate the engine forward with a wrench on the front pulley until the piston touches the stop. Mark this spot and rotate the engine backward until the stop prevents it from turning any further, and mark the next spot. Top dead center is located exactly halfway between the two spots on the crank pulley.

Check Injectors—If you have spark, and it happens at or near the right time, but the engine still isn't trying to start, the next step is to search for fuel problems beyond the earlier visual inspection and test of the fuel pump. Checking for a low or a complete lack of fuel pressure is pretty easy and a good first step. If fuel pressure checks out, but you suspect the injectors aren't delivering any fuel at all, then check with a volt meter to be sure the injectors are supplied with 12-volt power when you turn the key on. Next, crank the engine and see if the injectors are getting a pulse (a tool called a *noid light* helps here). You can also use a mechanic's stethoscope to confirm that the injectors are clicking. Then there's always the somewhat less scientific "pull a plug and smell for fuel" test. But, chances are your plugs would smell a bit like fuel anyway, so watch for false positives here. If they're actually still wet with fuel, it's a safe bet.

If the Injectors Are Not Firing—While it may be possible that injector malfunction is due to a defective ECU, it's much more common to have wiring issues or a bad setting in the ECU commanding the injectors not to fire. The wiring should be checked for continuity to the ECU with a multimeter, and to be sure the injectors are getting power. Exactly what settings may cause an ECU to give spark but not fuel vary from ECU to ECU, but the most common is a faulty sensor reading, causing the ECU to think you are attempting to clear a flooded engine and therefore shutting off the injectors as discussed earlier. It may also be possible that the cranking pulse width settings are so low the injectors are barely firing. Another possibility is in some cases, the ECU may be firing some injector channels but not all of them, depending on how it is configured. A sequential system could be set up for the wrong number of cylinders, or injector outputs could be miswired. You were supposed to use channels 1 through 4, and you wired up 3 through 6, and now only 3 and 4 are firing. Even a two-channel batch-fire system can support staged injection only bringing in the second bank of injectors above a set load or rpm.

Okay, so you've been through the above, you've got fuel and spark, all of your sensors are reading what appear to be normal readings, and the ignition timing is in the ballpark. Still won't start and stay running? Sometimes on a new installation, everything is physically working and the problem lies in the base map configuration. The ECU is commanding the injectors to fire, but it's either firing so little fuel that the engine doesn't catch or it is firing so much fuel that it floods the engine. One way to check for a flooded engine is to pull the spark plugs and check if they're wet with fuel. Because clearing a flooded engine can take time and is generally a pain in the tailpipe, you're often best setting your cranking settings to inject less fuel than you think you'll need, and incrementally adding more with each effort until it catches. If it starts and immediately dies, check your data log; chances are the cranking pulse widths were over and it had gone into "run mode," running off the map with the added fuel from after-start and warm-up enrichments, but it's likely these are not right yet and will not keep the engine running.

So at that point you've got your cranking pulse widths in the ballpark, but you need to work on the fueling after it starts. Usually when it starts and immediately dies you're short on fuel and need to add a bit. When you're first getting the engine started, the easiest place to add this is probably the fuel table, scaling up the area that the engine runs in immediately after it starts so that it will have more fuel. Once you get the car to stay running,

Check the injectors with a volt meter and be sure they're getting 12v power if you suspect they aren't firing.

you can properly dial-in this area of the table and then correct this problem the proper and permanent way, which is through adjusting your after-start and warm-up enrichments.

Checking Your Wiring

If you suspect a problem in your wiring, you can start by measuring the resistance from one end connector to the other with an ohmmeter, assuming of course that you're able to reach both ends. The resistance needs to be near zero, and you may also need to make sure it is not shorted to the ground or to other wires by checking the resistance from one end to the ground or to the circuit you suspect may be shorted. This can be a time-consuming process if you have several circuits to test, but there is no real shortcut here. Check for 12v power to your injectors in the crank and run positions of the key. Check to make sure your ECU has power in crank and run. Check the ignition coil and ignitor too. Check for spark by pulling a wire, putting a screwdriver in the plug wire, and laying it close to a ground while you turn the engine over to watch for the spark to jump to ground. (Keep yourself clear or you'll get a wake-up jolt.)

Check for Proper Grounding—Other than a break in your wiring, there are other problems that can be a bit sneakier to track down. This is a case where it's better to spend a little more time doing it right up front rather than come back later chasing a hard-to-find issue. Make sure you adequately ground your ECU, and ground it to the engine block/head according to the manufacturer recommendations. The ECU may consume very little power, but it sinks a lot more current to ground, as this is how it actuates nearly everything

it controls. Fuel injectors can draw 1 to 4 amps or more each. An idle valve can add a couple more. Add any additional accessories, boost control solenoids, cam timing control solenoids, etc and the current sinking back to ground through that ECU can grow pretty high. Ground it right and you shouldn't have problems.

Voltage Drops—This is a scenario where you have maybe 13v at one end of a wire, but less at the other end. It's going to occur if you've used wiring that's too small for the job. With undersized wiring, even though there is no load on the wire, the resistance may be very low. When you try to pull too much current for the wire, the resistance climbs, the wire heats up, and the component you are powering doesn't get the current it wants. Where are you most likely to see this? Well, anywhere you've used wiring that was too small to carry the current needed by the powered accessory. Starters are common. But in the performance world, coils are a commonly underwired component. People will upgrade their coil and use the stock wiring, which may work, but does not provide the coil with the current it really wants to charge to it's full potential. Aftermarket CDI ignition boxes are another potentially underwired component.

To check for this, put the multimeter in DC voltage mode, hook one end to the 12v source end (battery positive or fusebox) of the wire powering the accessory. Hook the other end to the end of the wire at the accessory side of things (at the coil/ignition box/whatever it is). Then your reading is going to show you the voltage drop that's occurring across that wire. With no load on the engine, it's likely to be 0v. Rev it up and see if the voltage differential climbs. To fully check this, do it under real load on a dyno. At high rpm it's not unusual to see that the coil is not getting all the power it really wants. And yes, you can use this trick on anything that consumes power. Is your fan getting the current it wants when it's on high? Inadequate wiring can cause components to perform poorly, or in some cases even fail. It can also cause the wiring itself to get warm or in extreme cases hot, and that has its own dangers.

Troubleshooting Your Fuel System

Fuel Pumps—While making sure the fuel pump does in fact pump fuel is simply a matter of wiring it to 12v power for a short length of time (as recommended earlier), making sure it is delivering enough fuel can be a bit trickier. If you suspect the fuel pump is not feeding your engine's needs, the best way to check is with a fuel pressure gauge that

Fuel lines tend to leak at the connections; leaks in the middle are considerably rarer unless the fuel line is rubbing against something it shouldn't be.

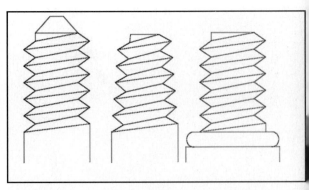

Fittings can seal in three different ways. Flared fittings seal against the edge of the cone. Pipe threads seal thread to thread. Straight threaded fittings seal with an O-ring at the end.

you can watch from inside the car, or a sensor that lets your ECU monitor fuel pressure. With a defective or inadequate fuel pump, the fuel pressure will usually start dropping when you ask the engine to deliver maximum power. Although a bad regulator can also cause the pressure to act strangely, it usually doesn't make the pressure fall off at full power. Sometimes the first symptom of an inadequate fuel pump is when tuning a car on a dyno and finding that you cannot seem to tune in enough fuel in the fuel table to enrichen the mixture to where you want it at high load. You richen up the table and the ECU is commanding the injectors to open for a longer pulse width, but there's no fuel to flow through them. Confirm this with a pressure gauge, and once confirmed, your only option is to upgrade the fuel pump.

Fuel Pressure Regulators—Fuel pressure regulators cause a different set of problems with fuel pressure. The usual symptom of a fuel pressure regulator that's too small in proportion to the rest of the fuel system is that it will deliver too much pressure at idle, and no amount of adjustment can bring it back down to normal levels. Signs of a regulator that's failing instead of merely undersized can include a lack in response to vacuum, which should reduce fuel pressure or, for turbocharged or supercharged applications, not responding to manifold pressure and failing to increase fuel pressure under boost.

Addressing Fuel Leaks—If you've used appropriate and safe fuel lines, it's very rare for leaks to form in the middle of a hose or line. If your system has fuel leaks when you pressurize it, most of the time they'll be coming from the fittings and connections. Each type of connection can have different problems.

Pipe threads such as NPT or BSTP fittings use the threads themselves to seal. They benefit from having a sealant applied to the threads. Teflon pipe tape can be used in a pinch, but can come loose and clog injectors, so it's best to apply sealant paste to the threads instead.

AN Straight Cut O-Ring fittings that thread into regulators and fuel rails may look a bit like pipe threads, but their threads just hold the part together and don't hold back the fuel. Instead, these seal with an O-ring at the top of the threads. This also applies to any factory fitting that uses a bolt thread rather than a pipe thread. If you have tried to use an OEM fuel pump with homemade lines, and you find that you can thread a bolt into the fitting where the factory fuel line goes, be sure your fitting seals against the top of the hole with an O-ring.

Flared AN or JIC fittings use a metal-to-metal seal with no rubber or sealant. They rely on a precise conical shape to prevent leaks. It's rare for AN or JIC hose fittings to leak, but if you are using hard metal tubing a leak from a flare fitting usually means you've formed the flare incorrectly. These sorts of fittings are very common and easy to work with though using stainless braided hose and readily available two-piece AN hose ends that simply screw on, making custom fuel hose assembly a snap.

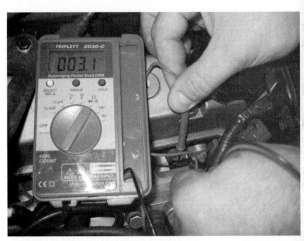

Measuring the resistance of an injector can tell if the coil is burned out, though it won't spot a plugged injector.

Checking an injector on a test bench.

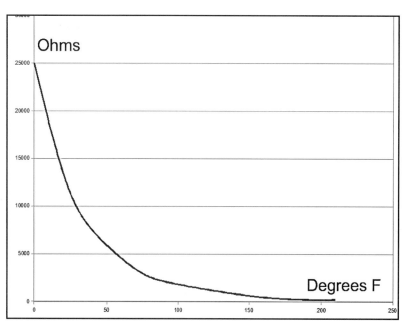

The resistance curve for the common GM temperature sensors.

Testing Fuel Injectors—There are a couple of basic tests you can do to a fuel injector with a multimeter. If you unplug the injector and measure the resistance across its two terminals, an injector with a damaged solenoid may read zero ohms or nearly infinite. The resistance you should get will depend on the injector type you have and you may need to compare it to a factory service manual or the manufacturer's data sheet, but most are either in the 2-4 ohm or 10-14 ohm range. Checking the harness side of the connector should give you 12 volts on the power supply of the terminal.

A mechanic's stethoscope is also handy for checking injectors, as you can hear them click when the injector fires. Sometimes, usually on larger injectors, the injectors are loud enough you can hear this without any help.

Testing to see if an injector is plugged requires somewhat more complicated equipment, but you can connect one to a pressurized supply of a safe test fluid and apply 12 volts to it to see if it opens. If your ECU has an injector test mode, you can even squirt it into a glass or graduated cylinder and measure how much fuel comes out in a given length of time.

If you've picked up a secondhand set of injectors and you're not sure how much wear and tear they've picked up, there are a number of mail order companies where you can send your injectors for a thorough cleaning and testing. A reputable injector service company will replace the worn items, run cleaning fluid through the injectors backwards (they force the fluid in through the nozzle and out the section where the fuel rail normally attaches), maybe use other methods like ultrasonic cleaning, and test the injectors' flow rates so you will know exactly how much fuel they'll give you.

Checking a sensor's resistance with a pan of ice water. This can then be heated to check the entire curve.

Checking Sensors

Temperature Sensors—Nearly all temperature sensors you see are negative temperature coefficient thermistors. This means that they are resistors that give a lot of resistance when they're cold and less when they are warm. If you suspect you're getting wacky temperature sensor readings, you'll need to set your multimeter to ohms and check their measurements. A service manual for your car will usually have a typical resistance versus temperature curve for your sensors. Shown above is a curve for the ubiquitous GM sensors that many aftermarket ECUs use.

If your thermometer tells you your garage is at 80 degrees, for example, you can check the curve and see your sensor should have a resistance of around 2000 ohms. You can also pull out the sensors and check them in a cup of ice water to get a handy

If your ECU allows you to calibrate it for different sensors and your readings look to be off, double-check your calibration settings and make sure you have the right ones for your sensors.

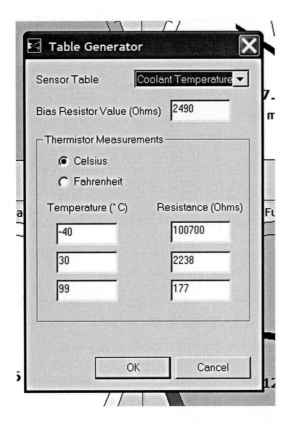

near 32-degree F reading. If the sensors are reading okay when you check them at their own connector, try measuring the resistance between the sensor wire and a ground wire at the ECU side. (If your wiring harness has a separate ground wire for the sensors, be sure you use that wire and not a different ground!)

With aftermarket ECUs, you can often calibrate their settings for different sensors. The procedure for how to do this varies from ECU to ECU. If your sensor checks out fine with an ohmmeter, but your laptop is showing the wrong temperature readings, try recalibrating the ECU for your sensors and see if that clears things up.

Intake air temperature sensors have another potential issue. If you thread one into an intake manifold, it can get *heat soaked*—that is, it absorbs heat from the manifold through the walls of the sensor, and this causes the ECU to get a false hot air reading. A common symptom of this can be found in data logs where the air temperature takes an abrupt drop when you rev up the engine. If you have heat-soak problems, consider moving the IAT sensor upstream of the throttle. Alternately, you can tune around issues in the correction tables of some ECUs, but it's often simpler and simply better to not have the issue to tune around in the first place.

Throttle Position Sensors—Throttle position sensors also get checked with a multimeter set for ohms. Unplug the TPS, put the meter on two of its

terminals, and move the throttle from open to closed and back. Most TPSs behave in the same way. They will have a reference voltage, ground, and signal terminal.

The resistance between the reference voltage and ground terminal stays the same. The resistance between the ground and signal terminals is low when the throttle is closed and high when it's open. The resistance between the reference voltage and signal terminals is the opposite, high at closed throttle and low at open throttle.

Another test is to back-probe the voltage between the signal and ground pins with the ECU powered up and the TPS connected. The voltage on the signal wire will go up as you open the throttle.

Some ECUs require a specific resistance or voltage value as well, but most aftermarket systems can adjust their calibration for nearly any TPS. There are a few exceptions, most of which use the same resistance curve as GM sensors.

MAP Sensors—The first thing to check about the MAP sensor is to see if it is reading the normal atmospheric pressure for your area when the ignition is on and the engine is off. This reading may be off by a few percent depending on the sensor tolerances and even the weather, but it should be close.

With MAP sensors, the vacuum line to the sensor seems to fail more often than the sensors themselves. Check the line for leaks and make sure it is routed where they can't be pinched when the engine moves. Also, if the line is routed into the passenger compartment, you want to make sure it won't be pinched when your passenger grabs the panic handle and braces his feet on your floor mat or kick panel as you hurl the car around the track. You should also make sure they are in a location on the intake manifold that provides a clear signal. Some manifolds give you a choice of several ports, and a port further away from the runner entrances and the throttle body is usually a good spot. Teeing your MAP sensor signal off your power brake booster hose sometimes gives a strange reading briefly when you step on the brakes. The right location can cut down on trouble with the sensor.

You can check the signal from most MAP sensors using a voltmeter. The manufacturer will specify a voltage range it will put out at normal atmospheric pressure. Some Ford MAP sensors are an exception; these produce a frequency-based output.

Airflow Meters—The vane airflow meters used on Bosch L-Jetronic systems (and their many licensed copies) are similar to a TPS, and you can measure the sensor with an ohmmeter after taking the meter off the car. The hot wire mass airflow sensors put out a voltage that you will need to test with the engine running and idling. Often the manufacturer's service manual will mention what the voltage should be at idle. This value may not be all that useful if you have changed the engine's displacement or had to bump up the idle speed significantly, however.

Hall Effect and Optical Sensors—If you are wondering if your Hall effect or optical distributor is working, you can take it off the engine and spin it by hand. Leave the wiring plugged in and back-probe the signal wire with a volt meter. Depending on the design, it should either go from zero to 12 volts, or zero to 5 volts. If it stays at zero, you may have a defective sensor, or you may need to add a pull-up resistor. This may be built into the ECU, but if it is not, you can run a resistor from a voltage source to the signal wire as shown in the drawing to the right.

VR Sensors—The variable part in a variable reluctor makes them a bit harder to troubleshoot, as these sensors don't put out a steady voltage. In fact, with the engine off, they don't put out any voltage at all. If you have an oscilloscope, it's pretty easy to check one of these by graphing its voltage output. Without such a tool, you'll have to resort to more limited checks. The easiest check is to spin the trigger wheel and measure the sensor's output with an AC voltmeter. You should get a slight voltage that increases with rpm, although some VR sensors will put out more than 50 volts at full speed. With the engine off, you can also measure the sensor's resistance and compare it to the manufacturer's specifications for the sensor.

Oxygen Sensors—Narrowband oxygen sensors use a voltage-based output and are pretty straightforward to check with a voltmeter: They put out around 0.8 to 1.0 volts under a rich mixture and 0 to 0.1 volts under a lean one. Measuring the speed of their response, which gets worse as the sensor ages, is tougher. Factory ECUs often use a method called *cross-counts*, where they vary the air/fuel ratio closely around stoichiometric, to check this.

Wideband sensors are more complicated to check, with the diagnostics usually built into the control device. Usually the controller's manufacturer will have more detailed information on how to troubleshoot their systems, and it's often model specific. One trait all wideband controllers share, however, is that they are precision instruments and

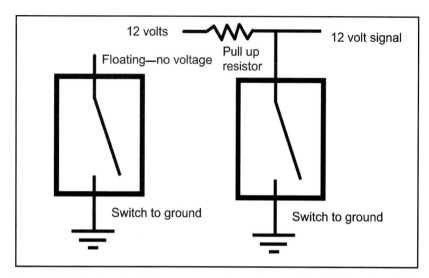

Some Hall effect and optical sensors need a pull-up resistor. This is often built into the ECU, but you can install one in the wiring if you need to.

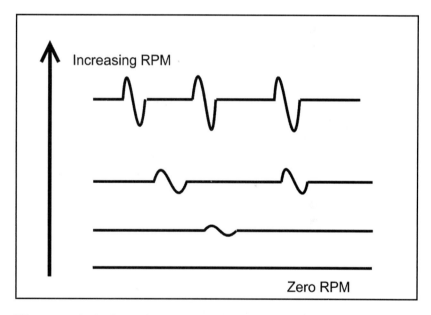

VR sensors don't give a signal unless the trigger wheel is spinning. You need to turn the distributor while checking the voltage from the sensor.

require precision grounding when it comes to wiring them up. The controller, if it isn't actually built into the ECU, needs to be grounded to the engine block/head and close to the same point where the ECU is grounded.

Chapter 14
Case Studies

We used this '77 Nova to compare a carbureted motor with low-budget junkyard parts to one with pricier but more sophisticated EFI options.

While we've tried to keep the whole book more practical than theoretical, this chapter gets to be especially practical—here's where we provide some details and near walkthroughs on EFI systems we've installed on real cars, and share the results with you. There's seldom one right way to build an EFI system; there are often a handful of right ways and dozens upon dozens of wrong ways. If you use an ECU with the feature set your engine needs, have a fuel system large enough to supply enough fuel, take the time to get it tuned and the bugs worked out, and have a big enough checking account to afford the system you've chosen, you'll probably end up with a right choice.

One choice you'll notice that all the examples in this chapter have in common is that they all use Bowling & Grippo MegaSquirt ECUs. We just happened to have stacks of them, forgive us. The principles are the same for most other standalone engine management systems currently available, although there are a few systems out there that would not support some of the features used on some project cars, as there are other top of the line systems with feature sets that we just wouldn't need to get into on these cars. These case studies are meant to help highlight some of the key differences you might run into installing an EMS system onto different vehicles, particularly vehicles with different types of ignition systems, since the ignition system is going to be the key component that makes the biggest difference from vehicle to vehicle when planning an EMS installation.

Some things are relatively constant. If you think about it, all temperature sensors are pretty much the same except slightly different calibration. All variable TPS sensors are as well. Narrowband O_2 sensors, yet again, not too much variation there. There may be a heated three-or four-wire sensor versus an older one-wire sensor that grounds itself through the pipe, but the bottom line is the ECU only cares about that signal wire and all narrowbands that put out 0-1v signals. And with injectors, again, the need to choose between high- versus low-impedance style injectors is the same, and that typically doesn't affect the wiring. The key subsystem that you'll be working to take control of that differs from engine to engine is the ignition system.

First, the ignition system differs on the input side, where the EMS gets its engine position/rpm signal from the crankshaft and/or camshaft position sensor(s). This trigger wheel arrangement can be mounted inside the distributor, in a separate module that turns at cam speed instead of the distributor typically called a cam angle sensor (CAS), or on a crank pulley or flywheel mounted wheel spinning at crank speed.

Second, the ignition system differs on its output side, the spark side of things. This can be anything from a single coil distributor to a coil-per-plug. In our case studies, here we wanted to show you a few examples that would help you see how the process of planning for each of these systems works. We cover distributor-based installs and wasted spark installs here, and we'll discuss the differences of planning for a coil-on-plug install. We also took the opportunity to choose a variety of vehicles that present many different EFI scenarios: a classic Chevy that never dreamed of EFI before; a factory EFI Toyota MR2 that allows for easy takeover of the factory wiring; and a full-on kit car that was completely built from scratch.

We chose standalone EMS systems for different reasons on the different cars. Using stock ECUs on these particular cars would have presented additional challenges: the MR2 has few options for tuning the stock ECU, and in our testing didn't respond properly to a piggyback computer we tried years ago. The Nova was originally carbureted, and the ECU for the kit car's engine donor does not lend itself easily to transplants. When it comes down to it we simply found the most flexibility with a standalone EMS.

Each of these cars has its own specific challenge that demanded changing its engine management to give us the tuneability we desired, and so with each car here we will focus on different aspects of the installation. These are more like snapshots of problems encountered and solved, parts tested, and the like than completed projects. Of course, whether a project car is ever complete is another question; as most of us have an insatiable need to continue to tweak our rides to be the best they can be. We'll continue to tweak these vehicles as well. If you'd like to see where these cars are now, future mods to them will be documented at DIYAutoTune.com.

Case Study: 1977 Chevrolet Nova

Starting Point—We built this Nova to explore different ways to add fuel injection to a carbureted motor, starting with a low-budget swap involving junkyard parts, and working our way up to more expensive aftermarket parts, to help people see if you really do get what you pay for when it comes to EFI. The car rolled into our shop with a 350 small block out of a '76 Malibu (the car was originally a straight-six model), with very little in the way of modifications. Before testing the EFI, we installed a set of headers, an Edelbrock Performer dual-plane manifold, and a Holley 600 cfm four-barrel carburetor. Pretty much your basic bolt ons. We did this to see how the car would perform with some basic mild upgrades to the induction system so we'd have something to compare our results to later.

The Nova had a mostly stock 350 under the hood when it arrived at our shop. A previous owner had added a couple chrome dress-up items, but little in the way of internal changes.

We chose to go about adding fuel injection to this car in stages. This approach would let us take control over the engine's operation one step at a time, letting us assess the effect of each individual change and document the process for others to follow. This can also make troubleshooting easier for a backyard mechanic. We tested out both throttle body and multiport injection, starting with a junkyard GM TBI system controlled by a MegaSquirt-II ECU.

As for the ignition system on this vehicle, it's a classic that you're more than likely familiar with. This engine ran a coil-in-cap four-pin HEI distributor with mechanical and vacuum advance. The trigger wheel itself is inside the base of the distributor and has eight teeth, one for each cylinder. This toothed wheel is read by a variable reluctor (VR) sensor and the resulting sine wave is passed to the HEI4 module, which then triggers a cap-mounted coil, with the rotor and cap distributing the spark to the appropriate cylinder. What we did initially is let this whole archaic but functional system continue to do its job and didn't touch it, other than installing an Accel performance recurve kit that swapped out some of the springs to change the advance curve. We simply triggered our EMS system from the negative terminal of the coil, using this for our tach/rpm signal and installed our EMS in a fuel-only arrangement.

Checking the Fueling—The first step on this car actually wasn't even adding fuel injection; it was fitting the car with sensors to see how well the carburetor was working. While a wideband oxygen sensor is useful for measuring the overall air/fuel ratio, installing eight of these and measuring the fueling at each cylinder would be impractically expensive. So we installed one wideband oxygen sensor and eight EGT probes, one EGT probe for each cylinder. While you can install EGT probes into cast iron manifolds, we took the opportunity

The EGT probes attach to the headers with weld-in compression fittings. The fitting in this picture is tack-welded in place and ready for a complete weld around the base.

There is one fitting for each pipe, about 1.5"–2" from the ports.

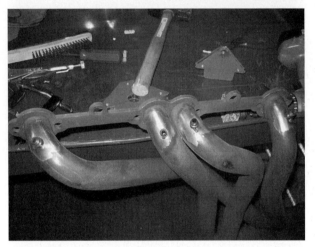

Having all the relays and fuses for EFI in a single module simplifies wiring and troubleshooting.

to also drop in a set of headers and build an exhaust system which could support future horsepower goals. Eight EGT probes might also qualify as impractically expensive for a junkyard EFI conversion; they're something we added to test the effectiveness of low dollar EFI conversions rather than a part of the conversion themselves. We installed two Innovate TC-4 thermocouple amplifiers and daisy chained them to an LC-1 wideband controller, allowing us to monitor both the air/fuel ratio and individual cylinder behavior through a laptop. The LC-1 also transmits its data to the ECU, while the TC-4s only report to the laptop. We made several dyno pulls logging the data from the carburetor, not just to get a baseline for how much power the engine made, but also to evaluate the fuel distribution and overall air/fuel ratio. The exhaust has an oxygen sensor bung on each side, letting us log the air/fuel ratio on one side with the onboard LC-1 and the other side with the wideband on the dyno.

Throttle Body Injection—Next, we wired up the car for a standalone ECU, a MegaSquirt-II in this case. The Nova did not have much in the way of room for adding more circuits to the stock fuse box, particularly not devices that need a relay such as a

fuel pump. The Bowling & Grippo line of products includes a relay board, a module intended for just this sort of installation. We mounted the relay board under the hood, while the more vulnerable ECU went inside the passenger compartment to protect it from the elements. A cable connects the two together. Since this cable has connectors at each end, we temporarily removed the backshell from the connector at one end to slip it through a smaller hole in the firewall. Since this car also used several other devices for data logging, we also added a marine fuse box to power the TC-4s and wideband controller.

The MegaSquirt-II uses an internal MAP sensor that connects to a vacuum tee. In addition to that, it uses an oxygen sensor, intake air temperature sensor, coolant temperature sensor, throttle position sensor, and a tach signal. Beyond the injector outputs, this particular MegaSquirt has outputs for a stepper idle air control motor, fuel pump, one channel of ignition control, and a general purpose on/off output. Our first round didn't use the ignition control output, so we wired the tach input to the negative terminal of the coil. With the exception of the oxygen sensor and fuel pump, all the inputs and outputs for the ECU happened to sit on the intake manifold, making it quite straightforward to construct a simple harness. We drilled and tapped the intake manifold for a coolant temperature sensor and welded a bung into the air cleaner for the intake air temperature sensor. Both of these were standard GM parts.

GM's throttle body injection systems are quite cheap to find in junkyards and easy to install on a wide variety of engines, so we started out with one of these. We bought a TBI unit off a 1992 Chevy van with a 350 small block for less than $35 including the injectors and wiring connectors. (Hint: Look for a K in the ninth digit of the VIN to be sure it's a 350 and not a 305. The 305 throttle body looks similar, but doesn't flow as well.) This included a throttle position sensor, IAC valve, and fuel pressure regulator built in. Even after having the injectors cleaned at WitchHunter Performance and getting a TBI adapter plate from TransDapt, it's a very cheap fuel delivery system. Thanks to adapter plates, it's fairly easy to put a TBI anywhere you can put a four-barrel Holley carburetor. We managed to adapt the Nova's stock throttle and kickdown linkage using a piece of scrap metal to relocate the bracket that originally supported the linkage.

The throttle body even accepts common air cleaners for four-barrel carburetors, although an air cleaner designed for a carburetor can pose a restriction. GM originally put a collar-like spacer in between the air cleaner and the throttle body. Leaving out this collar and putting a normal open element air cleaner on the motor certainly looked more like a carburetor, but it cost horsepower due to the way it forced the air to flow. We ended up combining the collar, which helped the airflow the way it wanted to, with an Edelbrock foam filter to get an air cleaner setup that cleared the hood without significantly hurting performance.

The TBI already has the injectors, fuel rail, and fuel pressure regulator built in. So it just needs an appropriate supply and return attached to it. We opted to use the original fuel line as a supply line. There is a fuel tank vent line that looks like a tempting point to use as a return, but on this particular car, the vent line didn't flow well enough to fit that purpose, so we plumbed a new line to serve as a return.

Fuel Supply—Since the Nova's fuel tank did not have any provision for a return line, we took the tank off the car to find an easy way to add one. This tank has the fuel pick up and sending unit mounted on the top, so we simply took the sending unit out, drilled a hole in the top, and attached a hose barb using pipe nuts and a little dab of gasket maker just to seal it up as we threaded it together. Dripping the fuel into the tank from above the fuel level is theoretically louder than a return at the bottom of the tank, but considering the car already had a set of 3" exhausts with rather loud Flowmaster mufflers, keeping it quiet was not a major priority. If that return is making any noise, we can't hear it.

For the supply part, we originally ran a Walbro GSL392 fuel pump, along with two filters. The

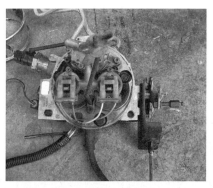

Our first EFI setup used this throttle body injection unit straight from the junkyard.

We used this adapter plate from TransDapt to fit the TBI onto a carbureted manifold.

The TBI unit, installed.

The carb air cleaner would not clear the hood, so we used this Edelbrock foam element.

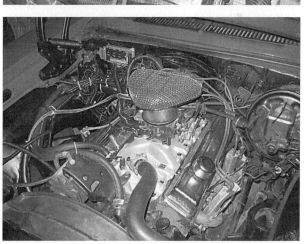

pump mounted to the rear subframe, ahead of the fuel tank, using two straps and sheet metal screws. While the car had a sock filter in the fuel inlet, it wasn't one we trusted to protect the fuel pump, so we used a regular parts store filter upstream of the pump, and a second filter upstream. The second filter catches any particles the first one misses (it has a smaller micron rating) and any possible particles that may come out of the pump from wear. For testing we used a small Russell racing filter, which

The fuel sending unit, drilled for a return fitting.

we eventually replaced with a larger SX Performance unit for long-term use. We were able to reuse much of the stock fuel supply line, plumbed to the throttle body with EFI-rated rubber hose. This isn't NHRA approved, but it worked for testing purposes. The entire fuel system is actually being replaced with all SX Performance components: fuel pump, both filters, and regulator, as we build this vehicle up toward new power goals. Both solutions work excellently depending on your goals. As we plan to replace the heads and cam, and add a turbo or two over the next few months, we're going to need to get some fuel up to the front (and shortly thereafter we expect to replace the shattered pieces of the short block with something stronger). Note that all of the pumps we used were multiport pressure capable pumps, but provided you have a regulator capable of bleeding off the extra fuel and maintaining the lower pressure needed for TBI, they will work just fine. This is why we started with the Walbro for the TBI setup; it worked great for us there and it provided a smooth transition to the MPFI setup handling medium level horsepower goals with out a hitch. For higher goals something along the lines of what SX Performance has to offer becomes a worthwhile investment.

Firing It Up—With the TBI plumbed and wired up, we first tested the system before trying to crank it. Powering up the fuel pump revealed a loose hose connection the first time we pressurized the fuel system, something that could have been quite dangerous if we had just attempted to start the engine and missed the leak. After correcting this, we confirmed that the sensors were reading as they should and it was ready to start cranking.

We found that the engine would start with a surprisingly wide variety of cranking pulse widths with the TBI in place. After tuning the system on the dyno, our testing revealed that it gave up a tiny bit of peak power, about 2 to 3 hp, but gained a bit

of low rpm torque compared to a Holley carburetor. It also gained a bit more power above the torque peak as the carburetor was running somewhat richer at high rpm. Fuel distribution as viewed with the individual EGT probes was no better than with the carburetor, requiring the engine to be tuned a little bit richer than would be ideal for gas mileage improvements. This prevented us from seeing any significant gains in fuel economy with only a fairly humble half mile per gallon increase over the stock configuration, but drivability and cold start behavior were much better than either of the two carburetors ever were after dialing the car in.

Ignition Control—The next step in tuning was to see what we could accomplish by giving the ECU control over the spark curve. So we went to the junkyard and brought back an eight-pin HEI distributor and coil off a late '80s GM van, and gave the distributor a new cap and rotor. We picked an external coil type distributor to give us flexibility with trying different intake manifolds, since some fuel injection setups won't clear the larger coil-in-cap distributor. We bolted the coil to the firewall.

This change in our ignition system required a bit of new planning in regards to how the ECU was setup. In reality it was minimal, but any change to how the ignition system triggers or fires will require you to revisit this at least for a moment, to make sure you're in good shape. In our case, the new HEI distributor we chose also has a similar eight-toothed wheel read by a similar VR sensor, but note the sine wave that's output for this VR sensor is passed on to the HEI module, which is designed for a computer-controlled ignition system. This interprets the sine wave and sends a suitable square wave signal out to the ECU to be processed. The ECU receives this signal, calculates the necessary ignition timing advance based on its inputs and programming, and returns another square wave back to the HEI module. This then fires the coil with the ignition timing advance commanded by the computer factored in. There is no mechanical or vacuum advance system on this distributor at all; the computer handles all of the timing needs with the exception of a limp home mode or base timing mode that's enabled when disconnecting a one-wire connector to the HEI module. That puts it in a special mode, where it ignores the computer-commanded timing. This can be useful when setting the base timing or to get you home if the ECU malfunctions.

Setting up the distributor required some minor reconfiguring on this particular ECU. We used a couple of pull off jumpers to reconfigure its input

The four-barrel Pro-Jection flows more than a junkyard TBI, but with a smog-era 350, the TBI's airflow is not what's holding it back.

With the chrome air cleaner in place on the Pro-Jection TBI, it's hard to tell the Nova is fuel injected.

circuit, and used the TunerStudio tuning software to change the necessary settings to let the ECU know it was now in control of the HEI ignition. It's important to change these settings before actually wiring the ignition module to the ECU, as the wrong settings can damage the module or coil by running too much current through it.

Because the HEI module can fire the coil without the ECU controlling it, we first fired up the engine with the MegaSquirt's spark output wire disconnected and set the timing to 10 degrees BTDC with a timing light. This ensures that the distributor rotor isn't too far away from the terminals when it fires the plugs. Next, we hooked up the spark output wire and adjusted the MegaSquirt's trigger angle setting so the timing in the tuning software matched what we saw with the timing light.

With the ECU in control of the timing, we

tweaked the spark advance curve on the dyno. We were able to pick back up the three horsepower we had lost since removing the carburetor. Not only that, but we picked up even more low rpm torque from a more aggressive advance curve coming on earlier than previously possible. In addition to testing computer control, we also conducted a short dyno showdown between the junkyard coil and an MSD HEI coil. The MSD coil did indeed put out more spark energy, as we verified with an oscilloscope, but that didn't translate into any more power at all in our case. This would be more of an appropriate mod on an engine that revved higher or made more cylinder pressure; our mild buildup just didn't need an incredible amount of ignition power to reliably ignite the mixture.

More Airflow—Another change we tried that may have been a case of right mod for the wrong engine was bolting on an older Holley Pro-Jection four-barrel 900 cfm throttle body. It looked a bit more like a traditional carb, and it's rated for nearly double the airflow of a stock TBI unit. However, the extra airflow didn't do anything for power, and fuel distribution did not improve either. It also lacked an IAC valve, something that later Holley throttle bodies corrected.

Going Multiport—This really left us with one more TBI option we could have explored but did not. That would have been modifying a stock or aftermarket TBI manifold to fit our Gen-I smallblock heads to test and see if air/fuel distribution improved. Instead, our next step was to install a multiport setup on our small block. We chose a Holley Stealth Ram, a two-piece, bolt-on intake manifold set up for port fuel injection. While it's possible to order the Stealth Ram manifold on its own, Holley also offers the Stealth Ram in a very comprehensive package, including the manifold, fuel rails, injectors, regulator, sensors, and throttle body. The Stealth Ram manifold itself resembles a Weiand Hi-Ram tunnel ram (in fact, when sold on its own, it has a Weiand part number). The lower intake closely resembles the tunnel ram except for having bosses for Bosch EV1 style injectors and fuel rails. The upper intake is a boxy plenum which accepts Tuned Port Injection style throttle bodies and linkages.

While the Stealth Ram intake is pretty much a direct bolt on to a classic 350, there are a couple parts that may not fit all small-block Chevys. We made plans to switch to a Stealth Ram early into the TBI installation, so we used a small cap distributor with an external coil. The Stealth Ram clears this distributor, but not the large cap HEI ignitions with the coil in the cap. Another fitment

The lower half of the Stealth Ram closely resembles a Weiand tunnel ram for a carbureted motor.

With the upper section on, the Stealth Ram starts to look more like an LT1 intake than a tunnel ram.

Spectre's chrome alternator bracket solves this problem by using a different mounting hole, one of the bolts holding the intake manifold to the head. This lets the same bracket work with quite a variety of intakes.

The upper half of the Stealth Ram has five ports. We switched around some of the connections a couple times before settling on putting the IAT sensor underneath the manifold and the brake booster fitting at the back. While these fittings look like they can be used for PCV, it's easier to use the fitting on the throttle body.

The Stealth Ram kit includes a throttle body originally designed for the TPI intake manifold. This lets us use a K&N filter for TPI applications as a simple air cleaner option.

It's a tight fit between the Stealth Ram plenum and the remote coil HEI distributor. An internal coil HEI distributor won't fit at all.

This hole in the manifold is used to secure the alternator bracket in place. Unfortunately, the Stealth Ram moves it about an inch forward, preventing us from using the original bracket.

issue was that it moved the thermostat hole and one of the mounts for the alternator bracket forward by about an inch. To bolt up our alternator, we replaced the top bracket with a Spectre Performance chrome bracket which used one of the manifold mounting holes in the cylinder head instead of the bolt next to the water pump, letting it fit a wide variety of manifolds. We also trimmed a couple inches off the radiator hose. For throttle and kickdown linkages, we used Lokar cable kits and a Lokar TPI cable mount. Removing the clevis from the throttle cable let us install it on the stock accelerator pedal. The kickdown cable bolts right up, although with the attachment point at the transmission being hidden behind the headers, actually getting it into place to bolt up takes some finesse.

Once fully tuned on the dyno, the Stealth Ram MPFI setup allowed the Nova to make a bit more power, while also improving fuel distribution and for the first time we saw a more measurable fuel economy increase of just 3.4 mpg over what we had seen with the stock carb setup, taking us from 14.3 mpg with the Quadrajet and stock cast iron intake to about 17.7 mpg with this latest setup. Better yet, this manifold should provide more airflow than the dual plane once we start building the motor for higher revs.

The original throttle cable was too short, not to mention worn after 30 years. We used a throttle and kickdown cable setup from Lokar, originally made for TPI motors.

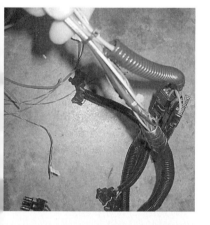

Piecing together a harness. The harness is held together inside using cable ties, and covered with convolute loom for a cleaner installation.

The thermostat location also required a little trimming on the upper radiator hose.

The Stealth Ram, installed complete except for the kickdown cable.

The MR2 is a great handler, but there's room for adding power.

Case Study: 1986 Toyota MR2

When Jerry turbocharged the 4AGE motor on his MR2 back in 2004, he knew the stock engine management couldn't account for the new airflow or properly manage the ignition needs of the turbo system. There is little that can be done to tweak the original ECU on these cars. He didn't want to change any more about the car's wiring than he had to, leaving as much of an OEM appearance as possible, at least when it came to the wiring (because there's nothing OEM-looking about a turbo in a first generation MR2). The challenge with this car was to convert it from the stock ECU to a standalone engine management system, while keeping as much of the original wiring as possible. While the installation didn't have to be plug and play, reusing known good wiring can make troubleshooting easier and make the modifications easily reversible. This install used a MegaSquirt-I, following in the footsteps of James Laughlin, a member of the online MR2 Owner's Club forums at MR2OC.com and the first person to have publicized such a swap back in early 2004.

Often, making a plug-in wiring harness is like one

To match up the signals, we'll assign each pin in the MR2 connector a number.

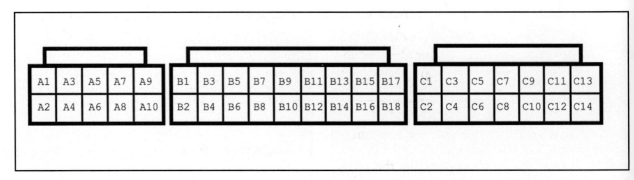

Stock ECU pin	Function
A1	Ground
A2	Ground
A3	Injector bank 2
A4	Injector bank 1
A5	Starter
A6	Ignition IGt
A7	Service connector
A8	Ground
A9	—
A10	TVIS (Variable intake system)
B1	Fuel pressure relief solenoid
B2	ISC valve
B3	Warning light
B4	—
B5	Service connector
B6	Coolant temperature switch
B7	Idle switch
B8	A/C input signal
B9	Ignition IGf
B10	Sensor ground
B11	G signal
B12	O_2 sensor
B13	G signal +
B14	TPS voltage reference
B15	—
B16	TPS signal
B17	Ignition NE signal
B18	Coolant temperature sensor
C1	—
C2	—
C3	—
C4	—
C5	Airflow meter
C6	Sensor ground
C7	Airflow meter
C8	—
C9	Air temperature sensor
C10	Vehicle speed sensor
C11	12 volt constant power
C12	—
C13	12 volt switched power
C14	12 volt switched power

of those match-up tests from grade school. You start with a pin-out of your stock engine control module and a pin-out of the aftermarket unit you will use to replace it with, and determine which pin on the stock module matches which pin on the aftermarket unit. Most of the connections are fairly straightforward. For example, 12-volt switched power, the coolant temperature sensor, intake temperature sensor, TPS signal, TPS VREF (5v usually), and ground are all easy to match up. Some signals arc a bit more complicated.

Ignition Setup—The most critical is the ignition setup, and here, the MegaSquirt and Toyota terminologies are a bit different, so a direct match isn't apparent and a bit of detective work comes into play. The Toyota VAST ignition system used on this MR2 has four signals that go between the distributor, ignition module, and the ECU, called NE, G, IGt, and IGf. The MegaSquirt only uses two such signals in a normal distributor installation. So the first thing to do is examine the Toyota documentation (or test the signals with an oscilloscope if the documentation doesn't give you all the necessary details) and see which signals are really needed and which can be dropped. The NE signal is a 12-volt square wave that gives one pulse per ignition event for a total of four teeth on the wheel on this four cylinder, so it's a natural choice to use as the MegaSquirt rpm/crankshaft position signal. The module itself actually switches like a set of breaker points from ground to a floating (not connected to anything) signal. So you also need to set up the MegaSquirt or other EMS so that it pulls the signal up to 12 volts when it's floating to provide the nice square wave from 0-12v that the MegaSquirt likes to see for crank position. The G signal gives one sine wave pulse every revolution of the cam. This would be useful for sequential injection and/or coil-on-plug ignition, where you need to know crank speed, as well as having a second wheel with one tooth indicating when cylinder one reaches TDC. This could be used if you were to convert a vehicle like this to wasted spark or coil-on-plug ignition, or sequential

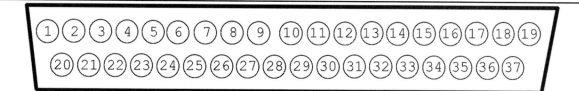

injection. The stock Toyota engine management system simply uses it to influence the order in which it fires two banks of injectors, but the particular MegaSquirt version used on this car does not need to be connected to this sensor. The IGt signal uses a 5-volt pulse to fire the ignition, so this hooks up to the MegaSquirt's ignition output pin and receives the commanded ignition output signal from the EMS, with the calculated timing factored in. Inside, the EMS needs to be set up for 5-volt ignition output. The IGf signal is an ignition diagnostic signal, and like the G signal, doesn't really need to be hooked up.

Matching Airflow Metering—The original Toyota engine management used a vane-style airflow meter. Besides posing a small but significant restriction, this meter runs out of its range when dealing with the extra air from a turbocharger. As the MegaSquirt is a speed density system, and has a MAP sensor built in, it's possible to remove the airflow meter. On this car, however, this presents a few challenges. The easier one is that the air temperature sensor is built into the airflow meter and not easily removed from it. The solution is to hack a connector out of a junkyard airflow meter (since this conversion was meant to be reversible, we didn't hack up the original) and wire a GM open element IAT sensor to this junkyard connector. The IAT sensor goes between the intercooler and the throttle body.

The other challenge is that the airflow meter controls the fuel pump. To keep the fuel pump from flowing with the engine off, the relay on this car wires to a switch in the airflow meter that turns it on when the meter reads a small amount of airflow. While you could rewire it so that turning on the ignition key turns on the fuel pump, this isn't the safest option in the event of an accident. The EMS has a fuel pump control output, but there's no obvious place to plug it into the harness as the stock ECU didn't control the fuel pump in this manner. However, you can take advantage of how both the ECU and the fuel pump relay are wired to the airflow meter. Wire the ECU's fuel pump output to the airflow meter's signal wire, and then you can add a jumper in the airflow meter connector to send this signal through this jumper in the AFM connector and straight to the relay

Megasquirt Pin	Function	Matching Toyota Pin
1-2	Ground	—
3-6	User defined	—
7-19	Ground	A1, A2, A8, B10, C6
20	Air temp. sensor	C9
21	Coolant temp. sensor	B18
22	TPS signal	B16
23	Oxygen sensor signal	B12
24	RPM signal	B17
25	User defined	A10 (optional)
26	TPS voltage reference	B14
27	User defined	—
28	12 volt switched	C13, C14
29	User defined	—
30	Fast idle	—
31	User defined	—
32	Injector bank 1	A4
33	Injector bank 1	A4
34	Injector bank 2	A3
35	Injector bank 2	A3
36	Ignition output	A6
37	Fuel pump	C5

controlling the fuel pump just like you want to, from the ECU.

Make the Wiring Harness—With the pin-outs matched up, the next step is to connect these terminals with actual wires instead of lines on a piece of paper. Jerry used a junkyard ECU and cut the connector out of it, then connected a short ECU harness to the appropriate pins on this connector. A couple tips can get you a professional looking harness without too much work. One, if you're soldering together a harness instead of putting the connector into a circuit board, there's no need to carefully desolder each pin out of the junkyard ECU. Just use diagonal cutters to trim the pins just above the level of the circuit board. Two, after soldering your wiring to the ECU pins a bit of heat shrink tubing over the solder joints makes everything a lot tidier, preventing potential shorts. Remember to slip the heat-shrink tubing on before you solder the wires; it's kind of tough to do after the fact.

Removing the stock ECU cleared the way for quite a few modifications that would never work

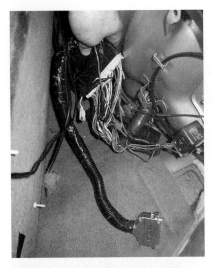

This adapter harness connects the MegaSquirt to where the stock ECU was.

The MegaSquirt itself sits in the trunk.

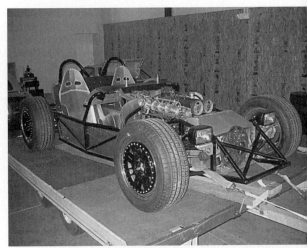

About a dozen Piper Engineering Seven frames have been built. This is currently the only one to use a Ford Zetec.

Case Study: Piper Engineering Seven

This is one of a short run of kit cars built by SCCA formula car designer Don Sievenpiper. The chassis was inspired by the minimalist Lotus Seven, a tube chassis covered with barely more than the sheet metal required to fill in the gaps between the space frame. Although Don originally planned the chassis around a GM pushrod V-6, car owner Roger Futrell wanted to use a high winding Ford Zetec out of a 2002 Focus. In stock trim, these motors made a rated 130 horsepower at the crank. Roger added a set of Jenvey throttle bodies to wake it up a bit more. Down the road, he's planning to swap in some more aggressive cams as well.

The Focus came with Ford's EEC-V engine management. This powerful controller tied into most aspects of the car's operation, including the instrument panel and transmission on automatic models. While cutting down on the number of

The MR2's engine compartment, crammed full of mods the stock ECU would not have supported.

with the stock ECU's tuning. The MR2 now sports a set of 440 cc/min Toyota Supra injectors, a T2 series turbo that has proven to be too small for the job really, Web cams, and a PWR air to water intercooler. This has brought the engine from it's lowly 100ish whp stock to a best of 169 whp currently, with everything being fully prepped to handle much more when we get the starter out of the way to make room for the much healthier GT2860RS turbo sitting in the shop right now.

The Zetec is a popular motor in kit cars, easy to find and with good aftermarket support.

The cowl on this body makes a good location for the relay board, fuse box, and wideband controller.

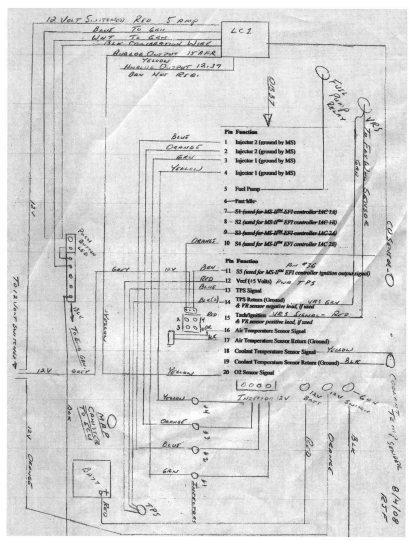

Wiring diagram for the EFI on Roger's Seven.

sensors the car used, lifting the entire set of electronics out of a donor car and copying every last detail of a Focus wiring harness is usually not practical on a kit car. Roger opted for a MegaSquirt-II to control his Zetec. While the Nova didn't start out wired for EFI, the Seven didn't start out wired up for anything at all. Roger opted to use the same sort of relay board used on the Nova, bolted to the cowl.

Independent throttle bodies can pose a couple of challenges for engine management, particularly when running open air horns without a plenum. Obviously, there's no good place to put a mass airflow sensor in an engine like this. Speed density systems can encounter some difficulty from the pressure waves in the runners affecting the MAP sensor reading, as well as the wildly quick transition from high vacuum to atmospheric pressure when the throttle is even lightly touched. This car ended up running alpha-N fuel metering, using the throttle position sensor instead of a MAP sensor to determine fueling and the spark curve. Another key sensor, the intake air temperature sensor, simply went into the air horn on the number four cylinder.

The Zetec has a distributorless wasted-spark ignition using a single trigger wheel spinning at crank speed. When dealing with this sort of engine, the first thing is to determine how the teeth are laid out on the trigger wheel. Like virtually all Ford four cylinder engines with a distributorless ignition, this trigger wheel was what is known as a 36–1 wheel, a wheel which would have 36 evenly spaced teeth except one tooth is missing, leaving 35 total teeth and a gap where the 36th would have been. With the engine set to top dead center, the missing tooth is nine teeth ahead of the crank angle sensor, or 90 degrees ahead. As a side note, having a single crank

speed missing-tooth wheel such as this is all you need for a wasted spark ignition system. For a coil-on-plug ignition system firing the coils sequentially or for sequential injection, you'll also need a wheel spinning at cam speed to provide cylinder identification. The crankshaft position sensor will tell you when a cylinder is at top dead center, but it won't tell you what stroke it is on.

Different ECUs handle decoding trigger wheels in different ways, and their tuning software usually reflects this. With the MS2/Extra software, you simply need to enter in the number of base teeth, the number of missing teeth, and the angle between the sensor and the first real tooth when the engine is at top dead center. In this case, since the missing tooth is supposed to be 90 degrees ahead, the first real tooth is 80 degrees ahead. Tests with a timing light revealed the real angle to be 78 degrees instead of 80, which the ECU easily accounts for.

Independent throttle bodies and high-performance cams can make for an unstable MAP signal when using speed density. We tuned this car using alpha-N fuel and ignition control.

Dealing with Mystery Electrical Interference—
Early testing revealed a problem reading the rpm while cranking. While the engine ran quite smoothly after starting, it took a long time cranking before it would catch on, with the rpm reading bouncing from 250 to zero and back in the tuning software. Our first reaction was to disconnect the coil and test the signal from the crankshaft position sensor. Using an oscilloscope showed a noisy signal from the crankshaft position sensor, and also indicated that the rpm itself was far from steady. One oscilloscope trace showed the rpm bouncing from 125 to 250 and back again twice in one revolution of the crankshaft. An effort at data logging individual signals to the ECU showed it interpreting the signal as one so noisy that it could not even pick out the missing tooth. If the ECU cannot reliably detect the missing tooth on the wheel, it cannot reliably determine engine position and rpm, and that is a problem.

Efforts to filter out the noise, both with hardware and software, yielded a few tiny improvements, but the cranking rpm refused to smooth out. It turns out this wasn't noise we should be trying to filter out, but it was noise that needed to be corrected at the source. Eventually we found more clues about the real problem. The battery drained unusually fast, even for a lightweight racing battery. We tried clamping the starter wires to a full sized battery in an effort to avoid destroying the car's tiny lightweight battery while sorting out the issue, only to have one of the wire's ends actually melt near where it was clamped. The starter itself became too hot to touch. We'd found the problem—a faulty starter with a shorting armature putting enormous amounts of noise into the electrical system as it drew enormous amounts of current each time it shorted out. Not to mention that it wasn't able to supply a good consistent cranking rpm. By the point it overheated, the starter was pretty much

dead. It may sound crazy to some that your starter can cause your engine not to start not because it's not turning over properly, but because of the electrical noise it's introducing, but at least in this case, it's the truth.

With a new starter in place, both the crankshaft position sensor signal and the tuning software's rpm signal smoothed right out. What seemed like a maddening, impossible issue to pin down turned out to be not really related to the engine management system at all. Once this was out of the way, the car fired up reliably and the ECU had no problem holding the timing steady throughout the rev range. We dialed in the base timing and setup the maps to begin tuning. With the fuel table dialed in reasonably well and the ignition table only roughed in with a conservative base table, we'd recorded a best of 98 whp. Then we started with the ignition table tuning, using the same methods outlined in the tuning chapter of this book. After spending some time working out the proper advance at low load and partial throttle angles, we moved on up to medium to fairly heavy loads at all rpms, then to tuning the wide-open throttle area of the table. As expected, we picked up a good bit of horsepower and torque dialing the ignition in properly, and landed at a fairly healthy 129 whp and 127 ft-lb of torque. Not bad at all for a 130 crank horsepower rated engine with ITBs and a good tune. And with a set of hot cams sitting on the shelf, I'm sure we'll have this thing back on the dyno again looking for more in the near future. At 1,400 lb and with an excellent suspension and finely tuned engine, this car will be a lot of fun when we get it out on the local tracks this season, as Roger has graciously offered us the opportunity to do. We're looking forward to it.

Taking It Easy

So that covers a few different types of installs and some of the hurdles we ran into that you might also come across or at least might be able to glean a bit of helpful info from. The best thing we can tell you when planning your installation is to do just that, plan the installation. Don't get overwhelmed. Some people, on their first EMS install, will try to dive in pull the EMS out of the box and just start wildly wiring things up without fully understanding anything about what they're doing. Maybe they'll come out the other side just fine. Or maybe they'll power it up and wonder why their laptop can't talk to the computer, or why the ECU won't seem to power up at all.

The method we typically recommend to a first-time EFIer to take on an EMS installation, and it's really not a bad idea for anyone, is to take it one wire at a time. You can take much of the mystery out of the install, give yourself little successes along the way to keep you motivated, and if you do run into trouble, you'll notice it immediately and know exactly what introduced that trouble. You just wired that single wire up and it couldn't be much else, now could it? So just wire up power and ground to your EMS. Hook your PC with tuning software up to it, or handheld programmer if that's what your system uses, and make sure you can communicate with the system. Don't expect any of the sensors to work properly just yet at all; they should all be displaying garbage, actually. Power it off now. Wire up the IAT sensor. Power it up and see if you've got a valid-looking IAT reading in your tuning software. You do? Cool. Power it off and wire up the CLT next. Test it. Wire the TPS next. Test it. Map sensor. Test. Etc, etc, etc. A plan like this is sure to help you get the job done, catch mistakes early, and get through the process with minimal frustration.

Happy tuning!

Checking timing on the Seven.

USEFUL FORMULAE

Fuel Pump Size
For gasoline, use these rules of thumb:

Liters per hour, naturally aspirated:
Fuel pump size = horsepower x 6.0

Liters per hour, forced induction:
Fuel pump size = horsepower x 7.3

Gallons per hour, naturally aspirated:
Fuel pump size = horsepower ÷ 10.5

Gallons per hour, forced induction:
Fuel pump size = horsepower ÷ 8.6

The flow rate is at the fuel pressure the engine needs at full throttle, which may not be the fuel pump's rated flow if it is rated at a different pressure. Fuel pump manufacturers typically provide charts of the pump's flow as a function of pressure.

Sizing a Throttle Body
Start by calculating the engine's airflow:

cfm = (maximum rpm x volumetric efficiency x cubic inches) ÷ 3456

If you can find a throttle body rated in cfm, a throttle body rated for 1.7 times the actual cfm should work well if the rating is at the common pressure drop of 1.5 inches of mercury (20 inches of water). This is the usual amount of pressure used to flow test four barrel carburetors, but it's more pressure drop than desirable. If throttle bodies for your engine are only rated by their size, plug the cfm into the following rule of thumb equation for the largest throttle body area that is not likely to hurt drivability:

maximum area = (cfm x 2.4) ÷ 300 ft per second

Ohm's Law
V=IR

V is the voltage, I is the current, and R is the resistance. Used for a lot of different electrical calculations.

Injector Flow Rate and Fuel Pressure
Changing the fuel pressure changes how much fuel your injectors will flow. You can calculate the change in flow rate with this equation:

injector flow rate = (target horsepower x BSFC) ÷ (duty cycle x no. of injectors)

actual injector flow = rated injector flow x

√(actual fuel pressure ÷ rated fuel pressure)

This equation is reasonably accurate as long as the injectors are within their recommended operating pressure range. Too much pressure will keep them from opening.

Calculating Dwell
Measure the inductance of the coil's primary circuit or get this number from the coil manufacturer, and measure the resistance of the coil's primary circuit with a multimeter. Once you have these numbers, you can calculate the dwell required. You will need a scientific calculator with a natural logarithm (ln) function. The equation is:

dwell time = -(coil inductance ÷ coil resistance) x ln [1 – (coil resistance x maximum coil current) ÷ (maximum alternator voltage)]

The voltage is the highest you will reasonable expect the alternator voltage to reach, and you can often find the maximum current from the specs on your ignition module or coil.

ABOUT THE AUTHORS

Jerry Hoffmann is the founder/CEO of DIYAutoTune.com. He discovered a lack of affordable engine management components and information when turbocharging his own car, and decided to combine his IT computer education with his hobby. His company specializes in equipping the do-it-yourself auto enthusiast with the knowledge and components to hand-build and implement EFI systems in their own garage. Over the years he has installed and tuned EFI systems on numerous vehicles, from street-driven classics and road course club racers to drag racers and land speed cars on the Bonneville Salt Flats. His goal is to help others to understand EFI in a practical hands-on manner, with less focus on the engineering-speak, and more on the practical need-to-know-how.

Matt Cramer is an engineer for DIYAutoTune.com and a member of the Society of Automotive Engineers. He began his interest in fuel injection after difficulties with a carbureted turbo setup on his 1966 Dodge Dart led him to explore more precise fuel control options. He is also a freelance automotive writer.

GENERAL MOTORS
Big-Block Chevy Engine Buildups: 978-1-55788-484-8/HP1484
Big-Block Chevy Performance: 978-1-55788-216-5/HP1216
Camaro Performance Handbook: 978-1-55788-057-4/HP1057
Camaro Restoration Handbook ('61–'81): 978-0-89586-375-1/HP758
Chevy LS1/LS6 Performance: 978-1-55788-407-7/HP1407
The Classic Chevy Truck Handbook: 978-1-55788-534-0/HP1534
How to Rebuild Big-Block Chevy Engines: 978-0-89586-175-7/HP755
How to Rebuild Big-Block Chevy Engines, 1991–2000: 978-1-55788-550-0/HP1550
How to Rebuild Small-Block Chevy LT-1/LT-4 Engines: 978-1-55788-393-3/HP1393
How to Rebuild Your Small-Block Chevy: 978-1-55788-029-1/HP1029
Powerglide Transmission Handbook: 978-1-55788-355-1/HP1355
Small-Block Chevy Engine Buildups: 978-1-55788-400-8/HP1400
Turbo Hydra-Matic 350 Handbook: 978-0-89586-051-4/HP511

FORD
Ford Engine Buildups: 978-1-55788-531-9/HP1531
Ford Windsor Small-Block Performance: 978-1-55788-558-6/HP1558
How to Build Small-Block Ford Racing Engines: 978-1-55788-536-2/HP1536
How to Rebuild Big-Block Ford Engines: 978-0-89586-070-5/HP708
How to Rebuild Ford V-8 Engines: 978-0-89586-036-1/HP36
How to Rebuild Small-Block Ford Engines: 978-0-912656-89-2/HP89
Mustang Restoration Handbook: 978-0-89586-402-4/HP029

MOPAR
Big-Block Mopar Performance: 978-1-55788-302-5/HP1302
How to Hot Rod Small-Block Mopar Engine, Revised: 978-1-55788-405-3/HP1405
How to Modify Your Jeep Chassis and Suspension For Off-Road: 978-1-55788-424-4/HP1424
How to Modify Your Mopar Magnum V8: 978-1-55788-473-2/HP1473
How to Rebuild and Modify Chrysler 426 Hemi Engines: 978-1-55788-525-8/HP1525
How to Rebuild Big-Block Mopar Engines: 978-1-55788-190-8/HP1190
How to Rebuild Small-Block Mopar Engines: 978-0-89586-128-5/HP83
How to Rebuild Your Mopar Magnum V8: 978-1-55788-431-5/HP1431
The Mopar Six-Pack Engine Handbook: 978-1-55788-528-9/HP1528
Torqueflite A-727 Transmission Handbook: 978-1-55788-399-5/HP1399

IMPORTS
Baja Bugs & Buggies: 978-0-89586-186-3/HP60
Honda/Acura Engine Performance: 978-1-55788-384-1/HP1384
How to Build Performance Nissan Sport Compacts, 1991–2006: 978-1-55788-541-8/HP1541
How to Hot Rod VW Engines: 978-0-91265-603-8/HP034
How to Rebuild Your VW Air-Cooled Engine: 978-0-89586-225-9/HP1225
Porsche 911 Performance: 978-1-55788-489-3/HP1489
Street Rotary: 978-1-55788-549-4/HP1549
Toyota MR2 Performance: 978-155788-553-1/HP1553

Xtreme Honda B-Series Engines: 978-1-55788-552-4/HP1552

HANDBOOKS
Auto Electrical Handbook: 978-0-89586-238-9/HP387
Auto Math Handbook: 978-1-55788-020-8/HP1020
Auto Upholstery & Interiors: 978-1-55788-265-3/HP1265
Custom Auto Wiring & Electrical: 978-1-55788-545-6/HP1545
Engine Builder's Handbook: 978-1-55788-245-5/HP1245
Engine Cooling Systems: 978-1-55788-425-1/HP1425
Fiberglass & Other Composite Materials: 978-1-55788-498-5/HP1498
High Performance Fasteners & Plumbing: 978-1-55788-523-4/HP1523
Metal Fabricator's Handbook: 978-0-89586-870-1/HP709
Paint & Body Handbook: 978-1-55788-082-6/HP1082
Practical Auto & Truck Restoration: 978-155788-547-0/HP1547
Pro Paint & Body: 978-1-55788-394-0/HP1394
Sheet Metal Handbook: 978-0-89586-757-5/HP575
Welder's Handbook, Revised: 978-1-55788-513-5

INDUCTION
Engine Airflow, 978-155788-537-1/HP1537
Holley 4150 & 4160 Carburetor Handbook: 978-0-89586-047-7/HP473
Holley Carbs, Manifolds & F.I.: 978-1-55788-052-9/HP1052
Rebuild & Powertune Carter/Edelbrock Carburetors: 978-155788-555-5/HP1555
Rochester Carburetors: 978-0-89586-301-0/HP014
Performance Fuel Injection Systems: 978-1-55788-557-9/HP1557
Turbochargers: 978-0-89586-135-1/HP49
Street Turbocharging: 978-1-55788-488-6/HP1488
The Engine Airflow Handbook: 978-1-55788-537-1/HP1537
Weber Carburetors: 978-0-89589-377-5/HP774

RACING & CHASSIS
Advanced Race Car Chassis Technology: 978-1-55788-562-3/HP562
Chassis Engineering: 978-1-55788-055-0/HP1055
4Wheel & Off-Road's Chassis & Suspension: 978-1-55788-406-0/HP1406
How to Make Your Car Handle: 978-1-91265-646-5/HP46
How to Build a Winning Drag Race Chassis & Suspension:
The Race Car Chassis: 978-1-55788-540-1/HP1540
The Racing Engine Builder's Handbook: 978-1-55788-492-3/HP1492

STREET RODS
Street Rodder magazine's Chassis & Suspension Handbook: 978-1-55788-346-9/HP1346
Street Rodder's Handbook, Revised: 978-1-55788-409-1/HP1409